D1532458

Dimensional Color

Design Science Collection

Series Editor
Arthur L. Loeb
Department of Visual and Environmental Studies
Carpenter Center for the Visual Arts
Harvard University

Amy C. Edmondson	*A Fuller Explanation: The Synergetic Geometry of R. Buckminster Fuller, 1987*
Marjorie Senechal and George Fleck (Eds.)	*Shaping Space: A Polyhedral Approach, 1988*
Judith Wechsler (Ed.)	*On Aesthetics in Science, 1988*
Lois Swirnoff	*Dimensional Color, 1989*
Arthur L. Loeb	*Concepts and Images, in preparation*

Lois Swirnoff

Dimensional Color

With 283 illustrations, mostly in color

A *Pro Scientia Viva* Title

B

Birkhäuser
Boston · Basel

Dimensional Color
Lois Swirnoff

Department of Art, Design, and Art History
University of California at Los Angeles
Los Angeles, California 90024/USA

CODEN: DSCOED

First printing, 1988

Library of Congress Cataloging-in-Publication Data
Swirnoff, Lois.
   Dimensional color.
   (Design science collection)
   "A Pro scientia viva title."
   Bibliography: p.
   Includes index.
   1. Color-Psychological aspects. 2. Visual
perception. 3. Color in architecture. 4. Environmental
psychology. I. Title. II. Series.
BF789.C7S95   1988   155.9'1145   88-6362

CIP-Titelaufnahme der Deutschen Bibliothek
Swirnoff, Lois:
Dimensional color / Lois Swirnoff: - Boston; Basel:
Birkhäuser, 1988
   (A pro scientia viva title) (Design science collection)
   ISBN 3-7643-3253-0 (Basel) Gb.
   ISBN 0-8176-3253-0 (Boston) Gb.

ISBN 0-8176-3253-0
     3-7643-3253-0

Typeset, printed, and bound by Universitätsdruckerei H. Stürtz AG, Würzburg, Federal Republic of
Germany.

By and for my students
and for
Joshua, Deborah, and Erica Gabrielle

Contents

Series Editor's Foreword

In a broad sense Design Science is the grammar of a language of images rather than of words. Modern communication techniques enable us to transmit and reconstitute images without the need of knowing a specific verbal sequential language such as the Morse code, or Hungarian. International traffic signs use international image symbols which are not specific to any particular verbal language. An image language differs from a verbal one in that the latter uses a linear string of symbols, whereas the former is multidimensional.

Architecturial renderings commonly show projections onto three mutually perpendicular planes, or consist of cross sections at different altitudes capable of being stacked and representing different floor plans. Such renderings make it difficult to imagine buildings comprising ramps and other features which disguise the separation between floors, and consequently limit the creative process of the architect. Analogously, we tend to analyze natural structures as if nature had used similar stacked renderings, rather than, for instance, a system of packed spheres, with the result that we fail to perceive the system of organization determining the form of such structures.

Perception is a complex process. Our senses record; they are analogous to audio or video devices. We cannot, however, claim that such devices perceive. Perception involves more than meets the eye: it involves processing and organization of recorded data. When we name an object, we actually name a concept: such words as *octahedron, collage, tessellation, dome,* each designate a wide variety of objects sharing certain characteristics. When we devise ways of transforming an octahedron, or determine whether a given shape will tessellate the plane, we make use of these characteristics, which constitute the grammar of structure.

The Design Science Collection concerns itself with various aspects of this grammar. The basic parameters of structure such as symmetry, connectivity, stability, shape, color, size, recur throughout these volumes. Their interactions are complex; together they generate such concepts as Fuller's and Snelson's tensegrity, Lois Swirnoff's modulation of surface through color, self-reference in the work of M. C. Escher, or the synergetic stability of ganged unstable polyhedra. All of these occupy some of the professionals concerned with the complexity of the space in which we live, and which we shape. The Design Science Collection is intended to inform a reasonably well educated but not highly specialized audience of these professional activities, and particularly to illustrate and to stimulate the interaction between the various disciplines involved in the exploration of our own three-dimensional, and in some instances more-dimensional spaces.

We experience our space through the interaction with surfaces. We may reflect light off these surfaces; we may touch or even deform these surfaces. If the region enclosed by these surfaces is at least partly transparent and the surface transmits light used in probing the enclosed region, then we may reach certain conclusions about that region.

A particular geometrical configuration may appear very different when its color, its context, or the color of the light used for observation is varied. We must therefore make a distinction between absolute reality and absolute truth. An absolute truth is almost always tautological and may be based on other, equally tautological truths. Mathematics is based on sets of postulates; if these are true, then truths may be derived from them. Mathematics aims at universality by reducing the number of basic postulates to a minimum.

Reality, on the other hand, tends to be subjective; our senses of touch and vision often give us mutually inconsistent clues. Because we can vary the apparent geometrical configuration of a three-dimensional structure by varying its color, we tend to separate form and color as if they were independent entities. Yet every form is observed through observations on its surface; when light is bounced off a surface, the result depends on the texture and color of the surface. Since surface cannot exist without texture or color, we cannot separate form and color using our visual sense alone. We can safely say that our conclusions about objects in our three-dimensional space are reached through a study of the properties of surfaces.

Lois Swirnoff was made aware of the relativity of the appearance of colors when she was a student of Josef Albers: the perceived color of a field is affected by the colors surrounding that field. Unlike Albers, Swirnoff works in true three-dimensional space. In three-dimensional space the perceived location of a colored field is influenced by its color. In her paintings, Swirnoff uses three-dimensional constructions as a constant foundation, upon which continuously shifting patterns of color, light, and shadow interplay and change. She uses these effects to modulate surfaces, to coax the senses into reinterpreting observations, and, by creating paradoxes and ambiguities, to caution us about the relativity of appearance. Swirnoff's use of sun and shadow for transforming surfaces at different times

of the day and in different seasons provides a significant new approach to surface in contemporary architecture.

In *Dimensional Color* Lois Swirnoff articulates the experiments which she conducts with her students in order to formulate a grammar for the interaction between light and surface. Although these experiments may be of interest as well to psychologists of perception, and the artist in turn will draw upon perception psychology, one should be aware of an important distinction. Whereas the psychologist designs experiments to isolate and to evaluate quantitatively certain distinct phenomena, the artist creates very complex systems in which these phenomena interact, mix nonlinearly, and cannot be separated and isolated. What is of prime importance to artists and to architects is what happens, not why. In her Aspen creation and in her Saratoga Springs stairwell, Swirnoff needed to know how differently colored modules interact, not why the eye or the brain processes the signals which they receive from these modules in the way they do. *Dimensional Color*, then, is no more and no less than the articulation by a significant artist of the method and system which she has developed to arrive at a desired result. Although this sounds simple, few artists are able to be as articulate.

Cambridge,   ARTHUR L. LOEB
Massachusetts

Preface

The experimental work presented by this book began as a painter's response to an achromatic environment. In 1956 color was eschewed by architects and the term "environmental design" had yet to be invented. Color as an influence on form and dimension were intuitive issues, excluded from modern architecture by the ethos of reductionism.

Expressive or evocative aspects of color were the province of painting, to be found in the suspended fields and deeply saturated hues of Rothko, or the dissonances of De Kooning's "Woman." The rational art of architecture concerned itself with the development of pure plastic relationships, with structure, tectonics, and form.

The context has changed. In the present color is used, as are historical and rhetorical elements; architecture approaches personal expression, and the boundaries between the plastic arts soften. The need for basic studies of color in three dimensions is greater now as the language of design becomes more eclectic.

The most relative factor in design, color is influenced by and in turn, influences its environment. While it is an aspect of surface its perceptual impact is not superficial. The interaction of color and form, space, structure, and pattern, as surface reflection of light, can be rationally combined and understood. These perceptually complex factors constitute visual experience. The visible world is colored. There is no achromatic form in nature, no more so than there is the experience of decontextualized visual "elements."

Implicitly in this study, color is considered an aspect of perceived dimension, as it is studied inclusively with form. As the arbiter of dimension, the eye measures things relatively. Rigorous visual analysis of colored models and the assessment of their effects can sensitize the observer to the possibilities of new visual relationships and structures. The aim is to understand complex interactions sufficiently to make them useful to the designer, and to provide a rational basis for the use of color in architecture and the environment.

Credit is due the several hundreds of students who have been my collaborators over the course of time. Often it was the question of an individual, "what would happen if ...?" which would open a new avenue or lead to the next question, making the study truly interactive.

Josef Albers, the great exemplar of color in this century, led with his studies in two dimensions, represented by his book, *The Interaction of Color*. His advice, "search, not research," is reflected throughout. Arthur L. Loeb encouraged me to gather together my investigations and confront the task of writing a book. His lucid commentaries and insights as editor have clarified the text and have been constructively critical and invaluably supportive. Lore Henlein, whose imprint, *Pro Scientia Viva*, appears on the title page, has contributed many and diverse efforts to the project.

My thanks to the Graham Foundation of Chicago for a grant to publish the color illustrations, and to the anonymous donor-participant for his generous contribution toward the reproduction of students' models. The UCLA Art Council is acknowledged for its contribution to obtaining the permissions. Kyoko Kunii clearly delineated the diagrams throughout the text. Thomas Michael Carr provided constant and invaluable assistance in the design and layout of the book. And the Yaddo Corporation awarded me the gift of time, which greatly enhanced completion of the manuscript.

Los Angeles   LOIS SWIRNOFF

Introduction

Color plays a significant role in the environment. The designer may recognize its expressiveness by using its associative attributes as embellishment or symbol. Rarely is it used to shape space, enhance or diminish volume, or assign position to an object in the visual field. While color has a constructive aspect, it is added as a last decision in architectural praxis, often subjectively or arbitrarily. Can color be rationalized as an aspect of architectonic space or form? Is it a primary constituent or aspect of structure at the outset of the design process? We deal with these issues here by presenting experiments with color and forms as basic "building blocks." Central to the issue is that hue and light influence appearance. Goethe observed,

> We now assert, extraordinary as it may in some degree appear, that the eye sees no form, in as much as light, shade, and color constitute that which to our vision distinguishes object from object, and the parts of objects from one another. From these three, light, shade, and color, we construct the visual world.... [1]

One hundred fifty years later this startling assertion contrasts with the modern assumption that form is "pure," while color is "mere sensation." This bias persists in design education, where generally each is studied independently of the other.

This study is based upon the idea that color is a constituent of form. It begins with the premise that color can be considered a dimension. As part of the process of visual organization, color components, hue and brightness appear to interact fundamentally with surface, volume, and space to influence their appearance.

The perceived location of colored planes or the grouping of volumes can be influenced, changed, or made ambiguous. The profoundly affecting environmental works of James Turrell are an example. Playing upon and displacing the perceptual clues of shape, he places colored light at precise angles to the corners of a large room. Dispersed by uniform colored light, the interior space is reconfigured as a large field of suspended color, visibly projected through its entry frame. Light and color become the subject and objective of the work. They inform and extend the sense of interior dimension.

Experiments with colored fields in space show that boundary changes can affect the perception of distance. Angular forms become more distinct or conversely, softer, when colored. Complex assemblages or volumes can be altered to appear simpler, and simple volumes complex, by the play of color on their surfaces. Concavity can exchange with convexity, and "illusion" displace reality, by control over patterns of light, shadow, and color. While Goethe's statement clarified the underlying abstract structure of illusionistic painting, the same formal constituents can affect the "real" or concrete environment.

Visual dimensions are relative and subject to their context. Because the most significant aspect of the environment is light, the experiments described in this book were done under concrete conditions of illumination. Some are table-top experiments in ambient studio light; others were constructed to be il-

luminated precisely by artificial light. Large-scale studies were undertaken in specific environments.

For the sake of clarity and focus the study has been limited to an investigation of surface. Color reflectivity, spatial configuring in depth, the interaction of colored planes in fields, angular relationships, and curved surfaces were studied with relationship to light, shadow, and color.

Since basic hues can define groups and visually connect or constellate, pattern may arise as a result of their recurrence in space. Geometric forms provided a basis for experimentation; from edges to planes to fields, from angle to edge, volumes to clusters.

Simple models yielded complex phenomena. Observations have been collective, and the models reproduced with consistent results. As the environment itself is visually perceived, and both the source and construct of the designer, our studies were assessed by eye. Surface reflectances can be measured of course, and since the models can be reconstructed, our experiments could provide a source of data. The discrepancies between what is physically presented and what is experienced however, may prove to be surprising.

The objective is to formulate a visual grammar, to provide perceptual building blocks upon which new visual structures can be based. An inclusive vision of the design process, the study replaces the approach to the "elements" of design. Systematic, but by no means exhaustive, it may open avenues of investigation and materials beyond the scope of those offered here.

While to the designer color dimension and its formal attribution may be novel, in nature color is integral and commonly associated with the experience of space and form. Blueness defines sky; its chromatic intensity is a sign of weather or geography. The patterning of color on a butterfly defines it. The byplay of boundaries, real or superimposed, is a natural strategy to reveal or conceal form. Color is surprising and ambiguous. While it enhances and varies the visual environment, paradoxically it clarifies it as well. Forms are made more distinctive and separated from their backgrounds by a change in color. At the same time, analogous hues can unify or integrate the visual field. Color conveys meaning by association, is expressive, and can underlie mood and symbol; it is a function of metaphor. However, visual metaphor depends upon visual logic. A poetic structure, to be apt, must be precise.

When reductive logic excludes the sensory, the result can be orderly, but sterile design. On the other hand, the unlimited presence of stimuli produces visual chaos. The contemporary urban environment reflects both kinds of patterns. Perhaps it is time to reconcile the two tendencies by truly educating vision, and to base complex, yet lucid order upon clear seeing.

1 Architecture and the Significance of the Surface

While color appears to lie on the surface, it is not superficial. It signals a sense of space, or assigns form in perception, and connotes meaning by association.

In the arts, form is not "pure" but is achieved concretely, unless the art form is purely conceptual. The surface is significant, because the handling of the material can reveal volume and space if it is well wrought. Formal appearance and expressive intention are closely related. Brancusi's "Bird" is memorable for the fluidity and character of its bronze volume. Mies' insistence on the articulation of edges set a standard for the use of steel.

In traditional Japanese architecture natural surfaces are regarded with great respect for their visual qualities and are worked by craftsmen for their subtle coloration and distinctive texture. Related to the environmental context from which they arise, these surfaces articulate and reflect the buildings' tectonics as well. Form, surface, color, and texture are integrally considered in Japanese architectural traditions.

Conceptually, aspects of surface were of primary importance to Renaissance form. The relationship of the human body to the environment was repeatedly and clearly asserted on the building surface, and abstractly represented in environmental space. The *braccia*, an arm's length, was a unit of measure in fifteenth century Florence. Proportion and scale reflected human dimensions, and were developed in an architecture that included its environmental and spatial context.

The term *prospettiva* signifies prospect as well as perspective. In the context of twentieth-century functionalism, we regard the mechanics of perspective to be important,[2] or its relation to the science of optics, but perspective may originally have been devised as a metaphor.

1-1

Perspective defines the stance of observer in relation to place. This position determines the essential referents of the construction, so that the horizon line is defined by the individual's eye level, and the vanishing point corresponds to the observing eye. The picture plane and the visual field are surfaces conceptually derived, to account for the limitation of the choice of the observer's position. The system is unified by this choice, as it gave rise to the possibility of projecting three-dimensional "reality" to a two-dimensional field. Consequently also, the sizes and shapes of objects and the relationships between them could be rationalized and expressed pictorially. Implicitly, these relationships refer to the point of view of the human observer.

While Western perspective set the environment in direct relation with the human being, giving him the possibility of surveying it, and empowering him conceptually with its dominion, a limitation prevails. The visual field may not be a surface, but its spatial framework circumscribes an area that is bounded and contained. Following its invention perspective made possible navigational implements and space travel because the explorer could focus. It resulted in the discovery of the New World, while at the same time it defined the boundaries of the stage and the framework of the urban complex. That the three-dimensional environment can be geometrically projected to a surface plane is essential to its spatial conception.

The limitation of the static vantage point is self-evident. In time and in motion a dynamic, rather than fixed referent is required. The paradigm of perspective as a vehicle of expression as well as a mechanical device is inadequate to describe the extended environmental field. Beyond the boundaries of a self-contained referent, the picture plane, the sense of space in time can be expressed more as an opening of continuous spatial sequences or segments in motion. To experience the urban environment, or the sequence of progression through architectural space, one moves within and through the field, and the task of perception becomes far too complex to be resolved visually, as a single "take" on experience.

If the static point of reference is replaced as a sequence of vantage points, progressively moving as the observer moves, can the visual perception of the environment be said to be contained within a frame of reference? If there is a spatial alternative to define these differences in modalities and their meanings, what would it be? Does a spatial field become a series of surfaces, the projected geometries of individual "framings" of the visual world, like the individual frames of a motion picture camera, which taken together, simulate the experience of a moving image? Apprehended in time, is the spatial field a continuum, and if so, by what means can its dynamic organization be described?

The issue is concretely experienced by walking through a city. The street is a continuum in space and experienced by progressive motion. Buildings are observed as volumes, with intervals of spaces intermittently shifting between, in front of, or behind them, as one continues. Open urban spaces, such as the piazza or town square, occur as large-scale, distinctive breaks in the continuous pattern. Grid systems of streets regularize the rhythmic continuum, and provide breaks at their intersections. While streets are rarely designed now, these sequences are not orchestrated, but may be enhanced by codes with determine uniform height or plot size in historical locales. The most coherent places depend upon some kind of visual structure; continuity, pacing, recurrence, become the factors of organizations in time. In a dynamic, fluctuating frame of reference color can perform as a mark, focus, signal, or localizing aspect of visual organization.

1-2

The surfaces of old cities offer examples of color structure within the spatial-temporal frame described. While they need not be "designed" environments can reflect distinctive characteristics, by the customary use of particular ranges of color. The character or flavor of individual cities has much to do with their coloration, made distinctive by local palettes of indigenous pigments and materials which accrue with time. While the ambience of each place is unique, I find that color usage or its organization in the environment occurs in a similar way.

For example, in cities where the continuum of a street is defined by rows of connected facades, a contrast of color serves to distinguish property lines. In the Old Town of Stockholm, the plane of a wall is differentiated by a play of cool blue against warm putty. Swedish codes restrict the use of colors in this district, giving rise to a distinctive and internally consistent ambience. The recurrence and frequency of warm neutrals, steely blues, deep greens, and grayed ochres "work" in the deep shadows and with the pale contrasting light.

1-3

1-4

The same characteristics in the back streets of Venice are evident in the facades painted here, rather than inlaid with semiprecious stones, as they once were, on the Grand Canal. Rows of facades, paced and separated according to shifts between yellow and green ochre, are distinguished as individual dwellings. Lombard or Venetian red against umber, seen against a fully saturated cobalt blue sky, is characteristic. Venetian color, rich and sophisticated, recalls the palette of the High Renaissance; the taste prevails to the present in the environment itself.

1-5

1-7

1-6

In Mexico rows of adobe houses, one storey high and deeply colored, contrast with the saturated unfiltered blue of the sky at high altitude. In Oaxaca brilliant cobalt blue and primary red facades, juxtaposed, astringently define the place.

Collective decisions seem to create their own syntax, just as spoken languages distinguish one group from another. A street may appear unified in Sweden, harmonic in Italy, and dissonant in Mexico, but in each case color has played a major role in its visual organization.

The persistence of a color can localize or focus the eye, so that at a corner or intersection the spatial interruption is connected by its recurrence, In Jalapa, Vera Cruz, a row of houses painted primary yellow and blue appears repeated after the intersection at the opposite corner, serving to bridge the interval.

1-8

1-9

1-10

Integration of another kind can be observed in a piazza in Parma. At its corner, a building painted a warm earth color opens the vista, contrasting with a more distant building in the enclosure painted gray-green ochre. In the center of the complex the cathedral, its stone facade a desaturated light yellow ochre, appears to be bounded by the two adjacent fields of facades. As one progresses through the piazza, arrangements of these three planes, shifting in relation to the moving eye, appear as larger and smaller areas of lateral surfaces. At one moment, the juxtaposed effect of colors in the buildings adjacent to it cause the cathedral facade to appear transparent; its color is an intermediary between those of the two surrounding buildings. Thus the spatial order in which the buildings of the piazza are arranged is contrasted by the ambiguity of the effect of the color of its facades. Whether this effect is planned or accidental, the limitations of a local palette enhance the possibility of these juxtapositions.

1-11

1-12

1-13

Orientation and position by color is another characteristic of vernacular usage. Front to side is stated by the contrast of hues on adjacent planes of the corner of a building. In Jalapa, bright pink makes a primary statement against pale blue, while a similar relationship is expressed with more subtlety in Stockholm, by the use of ochre and slate blue. Punctuating the space, a buttery yellow glows against richly saturated earth red, in a hidden alley in Venice, a complex interplay of surface, space, color, and light.

1-14

1-15

1-19

1-20

1-15

1-16

1-18

1-17

1-21

Apertures are also commonly and universally defined by a contrast of color. To differentiate them from the wall plane, windows and doors are colored. The distinction can be made between materials, a stone lintel, or a wooden frame, or by the painted surfaces of adobe. A function of economics, the visual statement is the same, colored boundaries articulate openings.

In the indigenous environment color is associated with form. As the surfaces of an urban place are rewrought in time, familiarity with its shape and with the continuous effects of ambient light may become assimilated in visual memory, so that the integration of color is a matter of reiteration or recall. The richness and subtlety of places redolent with color represents an accrual over time, rather than conscious design.

1-22

1-23

1-24

However, the integration of color with the environment becomes a matter of choice when the formal constituents are developed simultaneously. While color has been used in architecture, it has for the most part been superimposed. The decorative effects of hues used as detailing focus attention to the part. Italian Romanesque architecture did this to advantage.

The facade of San Miniato in Florence is covered with contrasting bands of whitish and green gray marble. Their distribution defines arches and doorways. The Romanesque plan, a hierarchy of nave and aisles, is suggested by the repetition of arches on the facade, expressing a relationship between interior space and exterior surface. The cathedral at Orvieto is entirely surfaced in its interior by contrasting horizontal bands of black and white. Surrounding the cylinders of columns, appearing flat against walls, delineating the curved surfaces of arches, alike in rhythm and scale, they activate by disrupting the architectural surface. At the same time, by their repetition as patterning, they seem to unify the spatial interior.

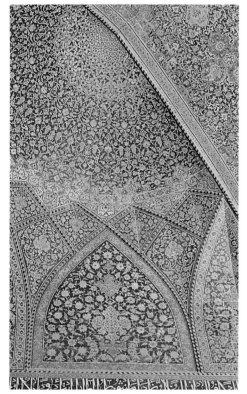

1-25

But the architecture most distinctly characterized by its surfaces is that of Persia. In this tradition surfaces are devised to draw the eye by the primary formal constituents, light and color, transcribed by mathematical precision to pattern. Architecture is expressed inversely, by its planes.

While from a distance a Persian building is singular and distinct as an architectural mass, relatively simple in shape and placed solidly in relation to its environment, at closer range the superimposed patterning of its planes replaces the solidity of mass with a fluid, changing order. An observer close to a surface can experience the changes in size and frequency of colored patterns of tiles and intarsia, as changes in space. Gradations of patterns and hierarchies of scale are devised to conform to volumetric masses, as they change their appearance. As a two-dimensional wall abuts a dome, the surface pattern will shift to adapt to the new topography. From the perimeter to the center of a dome, patterns contract progressively; a shrinking effect is emphasized by the remarkable integration of architectonic surface and imposed pattern.

Domes have been painted or colored by mosaic tesserae since their appearance in Byzantine traditions. The vaulted ceiling in Medieval painting and architecture was a metaphor for heaven, which was represented abstractly through coloration by gold leaf, or naturalistically, as a blue dome. By the High Renaissance, effects of diminishing distance were achieved by perspectival illusion, so that the dome as a surface ceased to exist, except as a visual reference to deep space. Replaced by imagery, of clouds, angels, putti, figurative narration, the Tiepoli transformed its surface to a pictorial field.

1-26

By comparison, the surfaces of early twentieth-century architecture appear mute or sterile. While the tenets of Modernism did not eschew color, its reductive ethos limited usage to primary hues. The transparent surface, the glass plane, served to reveal building tectonics, and a preference for structure predominated in modern architecture. Buildings were organisms in which the skeleton was more significant than the skin.

In Persian architecture colored patterns may be distinguished in their sequencing either as contrasts or as mixtures. From a distance the fine scale of a pattern will assimilate, so that the surface disintegrates altogether, like a hallucination, or meshes as a film, suspended in space. The patterns on exterior domes, predominantly blue, will suddenly integrate with the sky, and the surface of the building dematerializes.

Surface, therefore, is a primary attribute of Persian architecture. The elaborate stalactite shapes of architraves and semidomes are sculptural, built to absorb and reflect light. Much of the tiling that constitutes patterns is also three-dimensional, adding to the complex assimilation of color, surface, and mass.

Determined by visual devices that significantly alter perception, the same object can have multiple appearances. Changing orders define the observer's position at intimate or remote distancing, accomplished by the relationship of pattern to building mass. As he or she moves, either physically, or by eye scan, the observer is included by direct perception, in the architecture. This appears to be an inversion of Western perspective, where the observer by establishing his or her position in space, surveys all relationship within range of the visual field simultaneously. Perspective rationalizes space by conceiving of form relatively, while Persian scaling continually redefines formal appearance.

As the role of surface is reassessed in architecture and design, a return to decorative revivalism need not be the only alternative. Art Deco, so much in evidence in contemporary design, is a symptom of the revival of interest in decoration per se. In the midtwenties, however, color was used integrally on the surfaces of New York skyscrapers. Ely Kahn, in particular, was responsible for the revival of Greek polychrome and glazed terra cotta. He imposed color gradations on buildings, making them darker at the base and lighter as they progressed upward, to blend with environmental light and shadow. In the thirties, city ordinances imposed limitations on the heights of Manhattan's buildings, to ensure the presence of daylight between them. This "drawback" induced the setback, which, by

1-27

1-28

1-29

the innovation of the pinnacle form, created the characteristic New York skyline. Modern urban form then, has been influenced by sensory constraints.

More recently, the role of color in the environment appeared as the dominant influence in the metropolitan region of Los Angeles. For the 1984 Olympic Games, the offices of Sussman/Prejza and Jerde associates were consigned with the task of converting the urban arena to a display of pageantry. Temporary structures, kiosks, venues, scaffolding, environmental graphics were all unified by the recurrence of a color palette. Termed "Festive Federalism," its combined effect, an eclectic combination of magenta, chrome green, deep yellow, vermilion, aqua, black, and white, expressed the cultural pluralism of the games themselves, as well as that of the city which hosted them. Elements of Mexican-American and Asiatic colors were incorporated, evoking a lively, celebratory spirit. Totally encompassing the vast urban area, the recurrence of the chromatic scheme unified all of its disparate elements, and demonstrated how significant color can be in the urban pattern.

2 Dimensional Color: Theoretical Background

The dimensions of color can be regarded as a function of vision itself. Fascinating as they are, theories of perception or an understanding of the physiology of the eye, are issues apart from those of visual invention. Central to this study is that the perception of color is integral with the perception of form. Therefore a brief discussion of some theories may illuminate the experiments that follow.

A Theory of Form

Gestalt psychologists like Kurt Koffka, concerned primarily with form, made its perception a fundamental law, and were vigorously opposed to the concept of "higher processes" in the brain which imply interpretation of data acquired through the senses. Gestalt theory professed that properties of size, shape, color, and localization must be regarded as different aspects of the same process of organization.[3] Koffka believed that this process occurred in the brain, presumably at the level of the visual cortex. He conceived of it as a process in a field, analogous to the visual field itself. The parts of the field, the contours of a form and its background, are united or separated by forces of attraction or repulsion similar to electromagnetic forces. When the eye sees a shape such as a cube, the drawing or outline is seen inescapably in three dimensions, not as twelve lines on a single plane. The third dimension is produced by the fact that rotation in this case is the simplest way of perceiving the figure. Field forces operating by the principle of least action, produce a good figure.

Before the isomorphic field force theory, the classical "Phi-phenomenon," named by Wertheimer, established phenomenal movement as an immediate experience. Two separate stimulus points of light, operated successively, will yield apparent movement. This gave rise to the concept of a physiological shortcut in the brain, which makes the process of perception a cortical, not retinal phenomenon. It explained how static patterns of light and shadow can cause the illusion of a moving image, and gave rise to the invention of the motion picture.

The most significant explanation for the perception of depth, stereopsis, represents the disparity perceived between the retinal images of each eye. Koffka hypothesized that field forces might underlie the motor-muscular action of this disparity in binocular vision.

> If one sets up apparent movement in the frontal plane, but interposes a line across the path of movement, the movement will be seen to pass beneath rather than through the obstruction, thus producing an experience of the third dimension ... . In stereoscopic vision... it seems that a process within the organism is organizing or filling in between stimulated points in such a way as to produce tri-dimensional space, and with it a fairly faithful reconstruction of the physical world.[4]

Gestalt theories of organization provided generalizations concerning perception, which have been adapted by design theorists.[5]

1. Interrelatedness and compounding
   Good figure (economy of form)
   Configuration
   Unit formation and segregation
   Proximity
   Similarity and equality
   Closure
2. Self-closedness and circularity
   Perceptual aggregates are defined units (spatial and temporal), in the non-Euclidian geometric sense
3. Space and time building
   Koffka regards environmental time as a spatial field of forces
4. Flexibility and transformation
   Transposition does not disrupt the aggregate. Musically, a change of key does not destroy the melody.
5. Establishment and persistence of constant relationships
   "Frame of reference"
   "Field theories"
   the perception of a magnitude or dimension in relation to a norm
   Color and size constancy.

For the perception of space or depth they defined certain characteristic visual clues:[6]

Linear perspective
Apparent size of objects whose real size is known
Relative apparent motion of objects as the observer moves head (motion parallax)
The covering of a far object by a near one (superimposition or overlapping)
Change in color of distant objects; loss of sharp outline and detail, "aerial perspective"
Degree of upward angular location of objects in a visual field
Relative brightness
Disparity of retinal binocular images (stereopsis)
Relation of light and shade
Convergence of the eyes on a fixed object (in inverse relation to distance)

The concept of the *gradient*[7] is useful to unify or organize data in the visual field. While the term is used by psychologists, physiologists, and physicists, it refers to degrees of change as grades or steps over a unit distance. When it is used as a visual description, attributes of size, density, or structure, grades of brightness (light to dark), or texture (fine to coarse), any change in a given visual dimension, can be discerned and measured by a gradient.

In the visual field the absence of a gradient would produce an unstructured homogeneous pattern, the experience of indeterminate distance. A surface, an element of two dimensions, can be defined by the consistency of its perceived microstructure or texture. Diminishing size and scale of natural forms, particularly those found in a patterned or textured environment (a pebbled beach), or devices such as dots, lines, or squares, will show three dimensions; the same surface, consistently reduced in size in diminishing distance, will create the effect of distance.

In fact, one theorist believes that the loss of a perceived microstructure in the environment probably has more to do with diminished effects of color and objects in space than aerial perspective. The intervention of particles of air through scattering is significant only at great distances. In the environmental field, the softening of contours and diminution of detail are more likely linked to the loss of a perceived surface or microstructure.[8]

Within the visual field, the condition for the perception of an edge, or bounded surface, consists of a transition. Segregation in the visual field can be caused by brightness contrast, or a change in luminosity. Two hues of equal brightness, on the other hand, cause the appearance of uniformity (Liebmann effect.)[9]

It has been established that brightness, and not form, is primary in perception. It requires twenty-five times as much light to perceive a form in a hazy, indistinct manner than it does to produce just noticeable light. The form reported most at this level was the triangle. The figure least confused with others with minimal light stimulus was the rectangle.[10]

The enclosure of a figure area by its ground cannot be apprehended without a separating boundary (figure/ground relationship).[11] So-called segregating colors, like red and yellow, are defined by having hard, distinct boundaries, in comparison with nonsegregating blues and greens, which by juxtaposition produce soft boundaries.

A visual boundary will also be produced by a change in density or texture. A sudden change in density produces an abrupt visual line, or a contour in space. Both a corner and a contour are lines in the visual field, but they contribute differently to the perception of depth in the visual world. The first is related to the slope of a gradient and help the object look solid, the second causes it to stand out against its background. An abrupt step in gradient causes depth at the contour, and to this James J. Gibson attributes the impression that we can "see" empty space. While the concept of the gradient is a device to organize the visual world as a visual field, Gibson's concept ceases to function usefully when it comes to the perception of color. He concludes:

> One might speculate that variations in hue or shading as such do not produce the same compelling impression of depth that gradients of texture, line, size, binocular disparity and motion produce, just because they are not related to physical depth by geometric laws as the latter are. Variations in hue and brightness can and do produce compelling experiences of outline, form and pattern in the two dimensions of extensity, but their correspondence to experiences of solidity, depth and distance is less precise.[12]

Thus psychologists leave open the question of the spatiality of color. Some studies show that an observed color appears to be influenced by area. Colorimetric matches were made for red,

green, and blue test fields in an equally bright, but neutral surround. As this area increased in size, there was a shift toward blue and an increase in brightness for all three test areas. This suggests that peripheral retinal elements sensitive to blue play a large part in the appearance of large areas.[13] Other studies show that colored light influences the judgment of dimensions. Experiments with depth-space under differing illuminants found that depth is underestimated, or seems lengthened, under these circumstances. A change in the intensity of light produced phenomenal shrinking and extension of the spaces. The lengths of sticks are estimated as longer and bigger under warm light than they appear to be under cool. Red light causes a box to seem heavier than it would under green.[14]

The phenomenologist David Katz characterized the appearance of color as having eight basic modes:[15]

Film (*Flächenfarben*)
Surface (*Oberflächenfarben*)
Bulk or volume (*Raumfarben*)
Transparent plane (*Durchsichtige Flächen*)
Mirrored colors (*Gespiegelte Farben*)
Luster (*Glanz*)
Luminosity (*Leuchten*)
Glow (*Glühen*)

The first three are the most significant in visual experience, differing from one another in spatial or tactile characteristics. "Surface" denotes a specified location on a visual plane in conjunction with an object, impenetrability, or having a distinct microstructure. Volume color has the quality of murky liquid, the sea, fog, colors that are not only themselves substantive or "voluminous," but through which as a medium, it is possible to see. The ultimate mode, to which surface and volume may be reduced, is "film." Familiar in experience as the appearance of the color of the sky, of objects located at a distance where their surface structure is imperceptible, or the subjective gray of prevision, film color is indefinitely localized. Homeless colors, they may assume a frontal, parallel orientation, or tend to assume the position of a surface through which they are observed. Katz also characterized the Newtonian spectral colors as film, the ultimate stuff of vision.

Mabel Martin stated further:

> If a color is localized in one plane we call it bi-dimensional; if it is located in more than one plane we call it tri-dimensional ... the former would have reference to the smoothness and impenetrability, the latter in softness, looseness and the invitation to penetrate ... . In either case, what is being described is not dimensionality in the geometric sense, but the pre-dimensional nature of visual quality.[16]

Color dimension in the sense of its extension or space filling characteristic was recognized by Katz, but he concluded:

> Neither the color value of a single object in the visual field, nor the color values of many objects presented together, can provide the basis for their observed orientation in space. On the contrary, the orientation of colors with reference to the observer always takes place in connection with the perception of certain spatial relationships.[17]

Not the first to have designated different modes to the visual appearance of color, Katz followed Goethe in this regard. In his Farbenlehre, the work he considered to be his "best beloved child," Goethe designates yellow and blue as the two "original hues," each one step respectively from lightness and darkness. His description of "physiological" colors, is related to surface, volume and film; they are called "Diobtrical," or colors in a medium.

Space may be filled in which there is an opaque occupation of space, or transparent, which is the first degree of the opposite step. Color in such a medium ... changes from light to dark ... the highest degree of light ... the sun ... is dazzling and colorless. Seen through a medium it appears to be yellow ... If darkness is seen through a semi-transparent medium, a blue color appears ... this becomes lighter as the density of the medium is increased ... darker and deeper the more transparent the medium becomes.[18]

The Paradox of Color

The experience of color, upon which phenomenology and art are both based, seems to present two contradictory, but equally compelling faces: its constancy in appearance, and its tendency to change. As Josef Albers stated,

No color is conceived as what it actually is physically. Without special devices we never see a color singly, or by itself, as we may hear single tones, but only in relationship to the many factors which influence our vision, which transfer the optical (physiological) susception into a psychological effect (perception)[19]

It was Albers' conviction that "art was concerned with the discrepancy between physical fact and psychological effect." [20]

The phenomenon knows as "simultaneous contrast" requires, as its name implies, the direct adjacency of color and was first so named by M.E. Chevreul.[21] He observed the activity between two grays of different lightness, which, when juxtaposed, intensified the lightness and darkness of the other respectively. Adjacent colors affect one another also in regard to hue. There is a strong tendency for the eye to subtract the hue adjacent to a color juxtaposed with another, so for example, an orange adjacent to a red will appear brighter and to contain yellow, while the red, by contrast, will tend to look cooler, darker, and bluish. By placing a small sample of color within a larger field, the contrast effect will be intensified. When the area influenced in this way is relatively desaturated, achromatic gray, for example, the hue component of the field color will influence it. A small gray square in a red field will thus look green.

Albers' theoretical work and his painting were inseparable, and in the series, "Homage to the Square," which preoccupied him for more than twenty-five years, the reductive format enabled him to discover and invent an enormous number of spatial and dimensional effects, based upon the principle of simultaneous contrast. His book, *The Interaction of Color,* the fruits of his teaching, is an introduction to the perceptual instability of color, and the remarkable extent to which an artist's control over so-called illusory experiences, could formulate a visual grammar for color metaphor.

Color Constancy

Of greater interest to phenomenologists is the color constancy effect, or the tendency for colors to retain their identity despite considerable differences in environmental light. For example, an object may look white, but not bright, on an overcast day, or black, but not dark, in bright sunlight. Ewald Hering called the ability of the eye to detect and adjust to the effects of illumination on color, simultaneous or instant adaptation.[22]

Katz thought that color constancy depended upon the primacy of the visual field, or its "total insistence." A high degree of total insistence, for example, produces the impression of strong illumination, while a low degree causes that of weak illumination.

The constancy phenomenon is favored when the observer is aware of the total visual field. Under these circumstances, the object can remain relatively independent of illumination change, since the eye is capable of recognizing colors under different levels of intensity.[23]

Spatial relations within the visual field are important, as well as intensity gradients, with regard to the contrast effect. Henneman found that a medium gray test object surrounded by darker objects would appear distinctly whitish. But if a much brighter surface were introduced, the original test object would look medium or dark gray.[24]

The Impressionist painters, particularly Seurat, were aware of both the constancy and contrast phenomena, and made reference to "real" experience by playing upon these contradictory visual reactions to color.

2-1

In Seurat's great painting, "La Grande Jatte,"[25] areas of landscape use mixtures of hues, greens, yellows, and blues, quantitatively distributed as small dots (pointillism), so that in some areas one of these color influences will predominate over another. For example, in an area at a reading distance sufficiently close to the canvas for the eye to discern the dots individually, a predominantly green field contains more blue dots, while in another adjacent to it, there are more yellows. When read at middle distance, these adjacent areas contrast with one another, the one appearing bluer and darker, the other brighter and warmer. The decrease or increase in perceived intensity is due to the assimilation[26] by the eye of the dots of blue or yellow, as they are mixed with green in their respective fields. At a greater distance from the picture surface, when the adjacent areas appear reduced in size, they are perceived increasingly contrasted as bluish or yellow-green fields. Still farther away, when the painting can be seen in its totality, a complex hierarchy of effect takes place. The blue-yellow, dark-light contrast becomes more active in the picture and causes the strong psychological experience of an illuminated landscape on a sunny day. The fluctuation between color intensities and their visual analogues as light and shadow, arrest the condition of a moment in time. That temporal phenomenon in nature is expressed directly, through the complex interaction of colors.

Physiological Optics

The study of the biological receptor organ, in its response to radiant energy, the eye, offers conflicting theories. Ewald Hering and Hermann von Helmholtz offer two different explanations for some of the effects discussed so far.

Hering[27] thinks that the retinal receptor cells are organized according to complementary pairs of colors. His theory, based upon the primacy of color as red, green, yellow, and blue, conceives of a dichotomy of substances in the retina, three that build up (anabolic process) under the influence of radiant energy, and three that break down (catabolic). The former process gives the sensation of white, yellow, and red, while the latter yields black, blue, and green. Receptor cells are paired accordingly, to be receptive to white/black, yellow/blue, and red/green stimuli, and are designed to activate the opposite process when stimulated by one hue of a pair. Hering's studies of the after-image and simultaneous contrast place these phenomena called the opponent/response system as physiological in nature.

Young and Helmholtz[28] proposed a three-color system, which finds correspondence in the retinal receptors. The primary receptors, according to this theory, respond to the primary sensations, red, green, and blue, so that a set of receptors sensitive to the long wavelength of the spectrum (red), one to the short end (blue), and one corresponding to the middle (green), exist in the retina.

Edwin Land's Retinex theory[29] postulates a single operational system, combining the effects of retina and cortex. The theory proposes that all of the receptors in the eye and brain sensitive to long wavelengths operate as a unit, and shifts the problem of color perception to that of brightness. In experiments with filters for long, middle, and short sections of the spectrum, the difference to the eye were those of luminosity changes. In order to separate them, three sets of receptors exist independently in the visual system for the three differing scales of brightness. An image in color is the result of a liaison between eye and brain, by the superimposition of the three sensitivity variables. Land's theory best explains the constancy phenomenon, in which a color is recognized by the eye independently of variations in the reflected luminosities of the object. A single object can yield three separate reflectances, and it is the comparison between them that confers the sense of color.

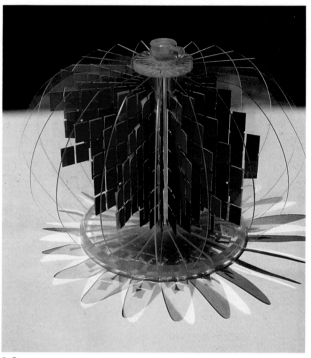

2-2

Under the influence of evolutionary theory, C. Ladd-Franklin[30] proposed a theory that accounts for the difference in function between the rods and cones. The former account for the response to acromatic sensation, the latter to color. Photoptic vision, or cone response to bright light, underlies the experience of saturated hue, while scotoptic, or rod vision, accounts for the achromatic grays seen in dim illumination or night vision. Ladd-Franklin regards the rods as the original mode of vision, and the physiological ancestors to the cones. She posits that, from an achromatic stage in the evolution of sensation, one set of receptors is to have split to provide yellow/blue sensation, and the ultimate development of the cones represents a further division of the yellow receptor cells to red and green.

The human eye is constructed so that the rods are distributed throughout the retina, while the cones are more numerous and densely packed in the central area, the region of the fovea. A small, circular pit in the center of the retina, the fovea contains about 250,000 hue-sensitive cells (cones) at a concentration of 160,000 cells per square millimeter, the size of the head of a pin. The macula surrounds the fovea. An oval body, containing rods and cones, it covers a visual angle of 3 degrees in the vertical plane, and 12 to 15 degrees in the horizontal.[31] Thus, a relatively small region of the eye serves the dual functions of sharpest focus and of the perception of hue.

2-3

Color Systems

The dimensions of color have been described by systems that model its variability, according to a variety of schemes. Those that are best known take three attributes, hue (red, blue, etc.), brightness (value), and saturation (chroma), and dispose them three-dimensionally.

The brightness component is geometric and can be described as the gradation of a scale from relative white to relative black. The Munsell system[32], shaped like a tree, has at its trunk or axis a vertical scale of values or gradations of brightness, from which hues radiate outward in a series of planes arranged around it as a circle. The distance of a given hue on this radial coordinate from the vertical axis of brightness, defines its saturation. Thus yellow, a comparatively bright (light) hue, at its maximum saturation (chroma), is located high on the vertical scale, nearer to white. The most saturated blue, a much darker color by comparison with yellow, is placed lower on the value scale, radially aligned with dark gray. The Munsell system is modeled asymmetrically, to account for the discrepancies between hue, saturation, and brightness.

The Ostwald system[33] is a variant on Munsell, which displaces the hues symmetrically on the radial axis, so that any hue at its maximum saturation is placed equally distant from the vertical scale of values. Unlike the Munsell system, which accounts for the differences in color variables, the symmetry of the Ostwald system belies visual experience.

Colorimetry systems measure the luminosity curves of the three variables for all colors, combining separate graphs into a total volume. Bouma's[34] volume, based on a Cartesian system, can be conceived as totally filled with points representing all possible hues, and their respective brightness and saturation. A given color point is completely enveloped by other points that touch it in all directions. Adjacent points along a given direction would represent differences in a given variable. Taken together, they form a closed surface around the point in question. The shape of this surface depends upon the spacings originally chosen to represent the three variables.

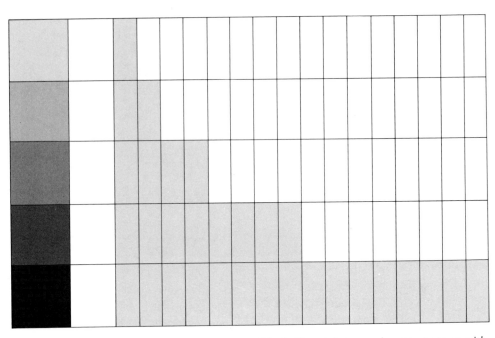

Fig. 1. Geometric progression appears as an arithmetic scale.

While these systems are elegant and represent a basic model for the complexity of color dimensions, none truly account for color perception. Brightness is a gradient, and can be expressed as a scale from white to black, with equal increments of gray between them, the sequence visualized arithmetically, as 1:2:3:4 steps of gray, actually represents a physical scale of geometric proportion, or $1:2:4:8^{35}$ etc., between intervals in light intensity, *Figure 1*. Thus, while light to dark is modeled misleadingly in color systems, it does represent an incremental change in a series, or a linear order.

Hue, on the other hand, is neither linear nor geometric, a fact belied by all systems. Yet Newton's R O Y G B I V, the spectral distribution of hues, discovered by breaking white light into its components by a prism, is represented by a circle.

The perception of color is not proportional to the wavelength that represents hue. For example, red, the longest visible wavelength, is relatively constant in appearance. By comparison, the shortest wavelength, violet, is highly unstable, readily influenced by mixture with its adjacent hues, blue and red. On a color circle, red and violet are juxtaposed and equally spaced. The hue orange, adjacent to red, appears distinctly divided from it, while blue merges with the hue adjacent to it, violet. Yellow, the third spectral hue, the most distinctive in its brightness, appears to be the most constant. It peaks in the spectrum, appearing more isolated from orange than from green. In the middle wavelength, blue and green readily merge and are highly interactive perceptually. Thus, while a color circle can model hues as adjacent to one another, a significant factor of mixture, the incremental changes as equal, sequential steps around a circle, is belied by their psychological appearance.

But the self-evident differences or individuality of hues is of greater significance. Of all possible choices, among a potential ten million distinctions between variables of light which can be distinguished by the human eye, how is it that precisely these seven are selected? Is it because they correspond to the source of the stimulus, light itself, in its spectral differentiation? Psychologically, these relatively few components, by their complex combination, give rise to the vast variety and subtlety of color in visual experience.

The tendency of hues to behave individually can be observed when they are juxtaposed and studied in context. A color interacting with its field will change appearance, but the apparent identities are unique in each case. A red can appear orange or bluish, depending upon the context by which it is influenced, and the degree of its saturation. A desaturated red in a brilliant red field can appear brown. Less saturated hues can change their appearance more

readily; as Albers noted, some colors are better "actors" than others. Highly unpredictable in context, as he remarked, "Color is the most relative factor in Art."[36]

The most obvious characteristic of color, the uniqueness of hue then, is of primary significance. It may be that the discrepancy between the linear, geometric character of light intensity and the distinctive, nonlinear attributes of hue, give rise to the experience of color as a dimension.

Dimensional Color

Imagine then, a visual world consisting entirely of color and light. Objects in, and the environment itself, would be distinguishable as distinct areas or patterns of light, shadow, and hue. The contrasts between them create bounded fields, which are perceived as shapes in depth. These shapes appear to be arbitrarily defined, for as the observer moves, or variations of light change, so do they. A three-dimensional volume is described as a shape determined by its relationship to a background or environment. And these are relative. Only a perfect sphere looks constant to a moving spectator, as a circle; and constant in size only when the distance between them remains the same. All other volumes change their shape, according to complex dynamics between boundaries caused by fluctuations of light, shade, and hue in the field and its background.

This is the universe of the artist, who structures its appearance with colored pigments in their magical association with light. If the tool of perspective is used these mobile data can be organized. As a system, perspective assumes a static position or vantage point, where the fields previously described as thresholds between areas of color and light, can be translated to line. The eye is an economist and performs a reduction. Monet or Bonnard may have preferred to maintain the original experience, by loosely defining the boundaries in an interplay of colors as shapes and fields, to correspond to the primary experience of the eye.

From this static position the shapes within the visual field assume a fixed appearance. They are clarified and ordered according to sequence and size. Completely bounded shapes read as forms that overlap incomplete ones. Similar objects can be immediately assessed by their relative size, and the depth of the field described by the degree of this difference. Straight lines group, so that, if they were parallel to one another, they appear to converge. If they converge away from the eye in a direction perpendicular to its axis, then these parallel lines seem to meet at a point directly opposed to the center of the eye, on a horizontal line.

The visual system organizes and reduces stimuli into a rational pattern. A coherent structure relates human experience with the environment and provides the implement for its understanding and manipulation. No wonder perspective, that artfully enduring device which converts "reality" to "illusion" has persisted!

But the Renaissance invention, in making a distinction between rationality and sense, *forma e colore,* caused their separation and development in independent directions. A reductionist dilemma now

2-4

persists; the distinction between form and color semantically shifts to the opposition between the original stuff of vision and the means by which it is organized.

It is the premise of this study that color is an aspect of form. If light is linear and hue nonlinear, then it may be that the discrepancy experienced between them by the visual system is the source of color dimension.

If we think of the environment as consisting entirely of reflected light, experienced as color, then spatial order would be a fit between a sensory pattern and its linear organization.

Aerial perspective was recognized in the Renaissance[37] by painters to be a device that closely approximates visible reality. The light of a distant landscape appears to have an orderly progression, independent of linear perspective. There is a tendency for the landscape to be perceived as layered or organized as a series of planes receding into space. In art these discrepancies have been defined as three planes in classical landscape painting, the foreground, middle-ground, and background of pictorial space. This

2-5

2-6

2-7

2-8

condition is so pervasive in vision that it has been abstractly represented by the juxtaposition of horizontal, uniformly colored bands. The abstract paintings of Rothko repeated this formulation, evoking comparisons by the banding of brooding horizontal fields of saturated color, with the mood and space of landscape. The condition can be repeated simply by the judicious choice and placement of solidly colored papers in a two-dimensional field.

2-9

Chinese scroll painting, which found its highest development in the landscapes of the Sung Dynasty, employed a similar formalism. The visual field, a vertical scroll, was divided into three basic sections, one at the lower edge describing a promontory or the beginning of ascent, a second, intermediary plane, containing a trail or waterfall, indicating progression along a path, and finally, a third, looming mountain form, darker above than below, terminated motion and expressed the relationship

with the sky. Landscape layers, developed as sequences in the field, show a contrasting band of the monochromatically inked edge, with a gradation to white scroll beneath. Distance is expressed as a progression from low to high in the field, and the contrast effect is clearly observed by the darkened boundaries of overlapping planes. The achromatic mode suggests a progression of light. The layering or alternate shifting from light to dark evokes transparency and is an effect that can be observed in distant space.

Meteorological optics defines this effect as the theory of visual range.[38] It explains the phenomenon as follows: Light entering the eye from a distant object is made up of two parts; one, the light of the object itself, which is *diminished* progressively by the process of scattering. The second, "air light," formed from the light of the sun and sky, has been scattered in the direction of the eye by the air intervening between the object and the observer. As the distance of the object increases, the amount of light reflected from its *surface decreases,* while the air light *increases* and approaches that of the sky. Thus, the color of surfaces at great distances appears desaturated, except for whites, which remain constant, and the light of the sky itself becomes an influence on the total pattern.

Consider this physical phenomenon in conjunction with the effect of one-point perspective. As the depth of the visual field increases in the perception of landscape space, the lateral range does so also. If the arc of centric vision is about 120 degrees, the lateral limitations of landscape expand with increasing distance. At one mile, the arc encompasses less of a horizontal field than it does at ten. How does the visual system deal with the increasing effects of light and space?

With respect to boundaries, the contrast effect diminishes between adjacent objects, due to the relative decrease in the difference in intensity between them.[39] But this change takes place within the context of an environment increasing in its radiant intensity or brightness to the level of the sky. When the eye reads distance as layers of illuminated planes, these differences are visible at their boundaries.

In these circumstances, what would be the function of color? In a completely undifferentiated visual field of light one can imagine the presence of the hue, RED. Making a pronounced statement a red would differentiate itself and fo-

cus attention in an otherwise homogeneous field. It initiates a sense of space. If a BLUE is introduced elsewhere in the field, the difference between it and the red would mark a separation of space. Redness and blueness then, would be primary distinctions to a responsive visual system, serving to differentiate what would otherwise be a luminous but empty field. The spatial denotation arrests the eye, as a place of fixation; the presence of more than one indicates direction. The tension engendered between two hues as distinct from one another as red and blue might evoke a physical comparison of interval and distance. As they virtually separate in the field, making distinctions, they clarify it as a space.

Then imagine a colorless universe. The orderliness of boundaries in the progression of luminous planes would yield an organized but dull experience. The primary presence of hues like red or blue, is an excitation that activates or sets up a discrepancy between a predictable or uniform organization and spatial action. I think that hue can be regarded as the sensory initiator of the experience of space.

In the chapters that follow, we will show how hue interacts with light, form, and space. Aspects of size, shape, distance, and field, the measurable dimensions of the visual world, are assessed by the eye, which integrates by experiencing them relatively.

3   Color–Space and Time

Hue is a sensory initiator of the experience of space. In the visual field bright hues arrest the eye, localizing as they isolate themselves from their surroundings. The aggressiveness or visual insistence of red is a spatial phenomenon, as is the radiating effect of a yellow, or the recessive quality of a light, desaturated blue.

In visual experiments with the spatiality of color, we sought an understanding of single hues, the combination of two and three, and then more complex sequences. Colors can be compared spatially only within an environmental field or frame of reference. To experience their spatiality hues may be assessed with relationship to other dimensional attributes, such as boundary or placement. In fact, the only example of unbounded color in visual experience may be the sky, and for most, even the sophisticated observers, psychologically, its homelessness is associated with the boundary of the earth. On a surface, or in space then, color remains relative to its context.

In his theoretical work, *The Interaction of Color,*[40] Albers simplifies the terms for the perceptual dimensions of color, reducing them to two factors, hue and light intensity. In his studies, the adjacencies of flat colored planes are perceptually active, in particular at their boundaries, where the relative intensity of the light and hue perceived in the contrast phenomenon mix with and influence the appearance of a colored field. These perceptual effects give rise to the illusion that a flat field appears transparent; in effect, the term *light intensity* is particularly relevant to perceived effects of color space and invoke psychophysiological comparisons.

As we were interested in the influence of distance separation at those boundaries, the series of experiments in color space used an open window in the form of a square. Through this formally neutral aperture, we could observe the behavior of colors behind it and separated in depth.

To what extent do colors perceived in space interact? Is this phenomenon dependent upon their organization in the visual field? Once colors are no longer perceived as a surface, but as film, will their separation in depth interfere with

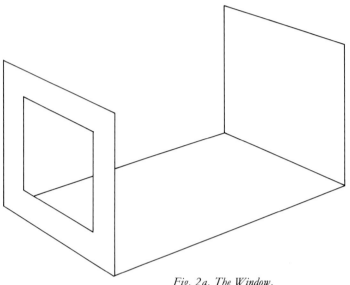

Fig. 2a. The Window.

boundary illusion or be enhanced by the induction of the contrast effect? Can there be color stereopsis? Is the sequence of colors significant, or does the eye organize patterns according to other spatial constraints? Do the subtle effects of radiation or expansion caused by the after-image depend upon two-dimensional juxtaposition, or can they also occur in a depth field? What is the influence of centrality and the relative sizes of edges and areas on the spatial perception of color?

The Experiments:
The Window Problem

In a series of experiments called the window problem, we worked with surface colors and incandescent light. An open box, containing a window, was devised to provide an individual frame of reference. *Figure 2a.* Experiments were done within the limited dimension of the box and observed and compared. Each student constructed an identical model of white mat board. Its dimensions were 6 inches by 9 inches in plan, and 6 inches by 6 inches in elevation. Placed in a frontal and parallel position to the observer, the 6 × 6 inch board was cut to show a 3-inch-square aperture. *Figure 2b.* Calling this the window, we looked at and through it to colors placed behind it vertically in parallel planes. Observations were made at a distance of 3 to 6 feet, both monocularly and binocularly.

The material used was Color-aid,[41] a silk-screened paper with a range of 24 hues, with "tints" (lighter axial values of hues mixed with white) and "shades" (darker axial values of hues mixed with dark or complementary col-ors), a total of 202 colors, and 18 intervals of gray, between a white and a black. The incandescent light, a flexible Luxo lamp, was incident from above or perpendicular to the 6 × 9 inch horizontal plane, to avoid cast shadows.

When colors are separated in depth and are observed singly through an aperture, they change in mode. While a sheet of Color–aid has microstructure that can be distinguished as a delicate texture, the silk-screened surface characteristic is transformed, when observed through the window, to the quality of a film. Having no microstructure to localize it, a film color tends to float, or to attach itself to the nearest boundary, according to a principle of "convenient visibility."[42] In this case, the boundary, defined by the window's aperture, localized the spatial experiences reported with color in the series of experiments.

3-1

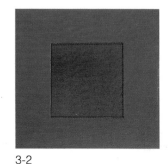
3-2

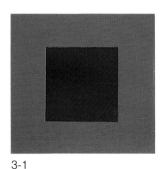

Fig. 2b. Window aperture.

Volume colors appear to be contained within the dimensions of the space-frame. Fully saturated, or deep colors, they seem to imbue, or fill the volume to the window. At the boundary they advanced slightly, as a viscous fluid attaches itself to the lip of a container.

Single Colors

Single hues were tested first, to estimate and visually compare their relative distances. A bright orange-red was chosen initially. The red paper placed against the back plane, 9 inches from the window, was observed as a film through the 3 inch aperture, which it appeared to fill. Within seconds the warm, brilliant red appeared to move from its spatial position within the frame, to float in front of the window opening. In this case the window as-

sumed the appearance of a square field, with the red superimposed or suspended in front of it. The change from surface to film color was perceived, and from a distance greater than 10 feet the spatial illusion was described as a projected light. After prolonged observation, between 30 and 60 seconds, the illusion intensified, and the red seemed to be located inches in front of the window. In no case did it appear in its true location, at the back plane of the space-frame.

Next comparisons between blues were made in a second space-frame. Observed at first separately, a fully saturated blue converted to film advanced to the level of the window, and after seconds "resolved" its location behind the frame. Placed adjacent to the frame containing the red, however, the saturated blue appeared appreciably more distant from the eye than the saturated red.

Noting the pronounced discrepancy between the color space of a red and a blue, we tested modified versions of each. After the bright orange-red we

3-3

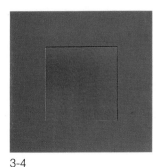

3-4

3-5

tried a darker, slightly blue red, the red hue of the Color–aid system. While the same displacement of true to perceived position was observed, the cooler red did not seem to advance as near to the eye as its warmer equivalent had previously. We tried a lighter blue than the fully saturated hue. When it was perceived as a film, its brightness increased, and its location, relative to the darker, saturated blue tested before, appeared to be closer to the eye, but behind the frame. When compared in their spatial appearance, the darker cool red and brighter blue were more equivalent in distance, the first in front of and the second behind the window. A series of desaturated (grayer) blues tended to recede. We found therefore, that the spatial behavior of reds and blues is a function of how intense or bright they appear to be by juxtapositions with the frame.

A dark, fully saturated blue appeared as a volume color to some observers. Mixed and desaturated hues of dark value, such as brown, also had a volumetric appearance. Deep violet was unstable as a volume color. If it contained

red, it tended to advance slightly; influenced by blue, on the other hand, it receded, in comparison with its blue or brown equivalents.

Brightness or value as well as hue conditions the effect of colors in the center field. In turn, a color seen through the window can be influenced by its interaction with the color of the frame. A fully saturated yellow, for example, tends to expand radially, rather than advancing, when observed adjacent to a surrounding white, bright window frame. When observed through a window covered by a dark gray, its brightness increases and the yellow appeared to advance. Colors, darker in value than the gray window frame placed before them, interacted by appearing brighter. The window behaves as a context then, much as a surrounding field would be, in a two-dimensional figure/ground relationship.

Bright hues tended to advance farther forward within the gray window than they had through the white one. Furthermore, a bright film color induces a darker contrast edge, and after fixed

3-8

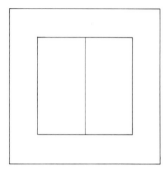

Fig. 3a. Apparent location in window.

3-6

3-7

and prolonged observation, it mixes with the gray at its adjacent boundary, causing a shadow effect, with an increased illusion of space.

When a hue appeared equivalent in its brightness to the gray window, it would appear to merge with it, or to share the same spatial plane (Liebmann effect). These experiences were reported also as projected films of light.

In all tests with single colors in the window, film colors changed their position relative to the boundaries of the aperture.

"True" relative to perceived location was dramatically demonstrated when, in a variation of this experiment, a large area of orange-red hue pieced together from several 18 × 24 inch sheets of Color–aid, was placed at the end wall of the studio and observed through a 3-inch-square aperture, held at arm's length. From a distance of 30 feet, the bright red, enclosed by the hand-held window, after some seconds appeared to move from the wall through the aperture, to float in front of it. The perceived distance was equivalent to what

had been observed before, when the red was 4 feet from the eye. *Note:* The large red field was uniformly illuminated against the wall. A surface in shadow, provided it is uniform, will behave as a darker red and will appear to move forward relatively less than its brighter equivalent. Film or volume colors assume positions in space, relative to the boundaries through which they appear.

Two Colors

To test the effects of two adjacent colors next, one was placed 9 inches from the window, at the rear plane of the frame, with the second placed parallel to it, and 3 inches in front. *Figure 3b.* Through the window the two appeared to divide the field in half vertically. *Figure 3a.*

The location of two hues is determined by the condition of their mutual boundary relative to the edge of the window. If they are equally bright, they may be observed as flat, and located in the same spatial plane. *Figure 3c.*

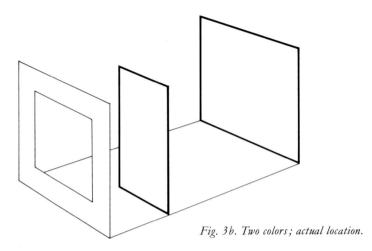

Fig. 3b. Two colors; actual location.

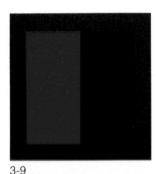

3-9

3-10

If one is bright, the other dark, the brighter color will assume a frontal location, parallel to and near the window, while the darker hue can appear to recede as a film. A hue, deep in saturation and dark in value may sink into the field, appearing as a volume color.

Two colors that are equal in brightness but different in hue, placed in two parallel and disparate sections of the space-frame, will assume the same spatial plane (Liebmann effect). In the case of two hues of near adjacency, a light blue and a light green, for example, the entire square field will assume the appearance of a subtle gradation between the hues. Brightness equivalence conditions the boundary. In a precise match none is visible. If this effect is observed through a gray window, close in intensity to the two test colors, the contrast effect between them is influenced by their relationship to the aperture. A hard boundary between window and field has the effect of softening the vertical boundary between the two test colors. Thus, in these close value relationships, boundaries are always relative to the total context. We next observed the same two colors adjacent to each

other against a two-dimensional plane, and experienced a greater contrast between them than was the case when they were perceived in space as films.

In effect, in almost no case are two colors precisely equal in intensity or brightness to the eye if they differ in hue. We tested the Liebmann effect with contrasting hues—that is, desaturated colors of near equivalence in brightness. A grayish red and grayish blue, equal in value were placed behind the window and observed. Whereas at first they appeared to be in the same spatial plane, in time they separated. As the contrast effect increased, the redder field tended to move forward, the blue to recede.

A vibration occurs when two hues of precise complementariness and equivalent brightness are placed in depth and observed through the window. In the window frame a bright, fully saturated red, and an equally aggressive bright green appear to be juxtaposed. They begin to interact immediately, both move forward and beyond the window aperture, and as the contrast effect intensifies, begin to "vibrate" at their

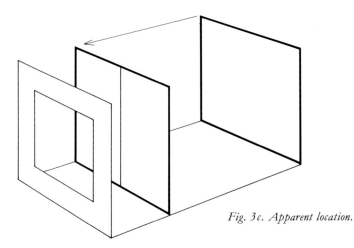

Fig. 3c. Apparent location.

3-12

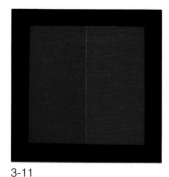

3-11

mutual edge. This effect also occurs if the complementary colors are studied adjacent to one another on a two-dimensional plane. The experience is spatial in either case; the mutual boundary appears to lift and quiver beyond the surface. In the case of the conversion within the window, the two hues appear suspended as films; in the two-dimensional reading the edge lifts above the field. Irrespective of "real" dimension, a visual depth dimension is obtained between two adjacent complementaries.

Sequences of Perceptual Mixtures

The floor of the space frame is separated by four grooves parallel to the window in three equal sections. This configuration was maintained for a series of experiments where four colors were observed interactively.

Using the window plane as one of the group, we selected four colors which related in their sequence as mixtures. For example, a bright, fully saturated green was the first choice, next to this was placed a less intense blue-green, next a darker green-blue, and finally a

fully saturated blue. They were related in a scale from bright to dark, and each represented a mixture of hues.

Placed in sequence on a flat plane, these uniformly colored papers will appear as gradations. The contrast effect at a mutual boundary, evokes an optical color so that for example, step 2 (blue-green) appears bluer, immediately adjacent to bright green, and greener, in its adjacency with step 3, the green-blue. A progressive gradation from green through blue converts four flat colors to their intermediaries in an illusion of mixture.

Careful to distinguish perceptual mixtures, attained by juxtaposition, from those resulting from the subtractive mixture of pigments, Josef Albers[43] had his students work with illusions of transparencies. If supported by an overlapping configuration, they could appear to be superimposed in space; or as an intersection, three flat colors, becoming two, as the mixture between is perceived as a transparency. In these studies, perceptually logical relationships between sequences of colors result as a spatial order in the eye. In three dimensions, the space is experienced as depth.

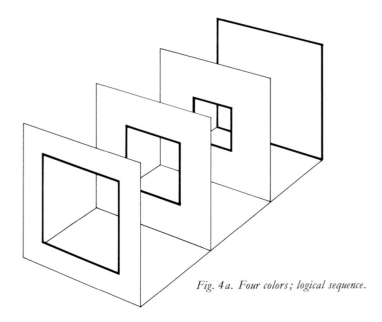

Fig. 4a. Four colors; logical sequence.

3-13

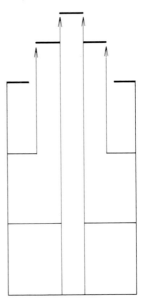

Fig. 4b. Perceived sequence.

Since the contrast effect activates illusions of transparency, in converting a flat surface to that of a gradation, distances between colors are estimated by observing the nature of their contrast boundary in juxtaposition with one another. A harder, or more pronounced boundary is obtained when the colors juxtaposed are more distant in their hue and brightness. Such a boundary appears to separate them spatially, as well. A softer boundary occurs by the juxtaposition of more equivalent colors, hence the apparent distance is decreased.

A range of colors can be selected by eye, by juxtaposing colored papers in an orderly sequence. A gradation from orange to red in four steps must correlate with a gradient in brightness. Finding adjacencies between more distant color ranges can become a visual game to the eye sensitized to boundary effects. If two colors are selected and intensely observed, the contrast at their mutual boundary will evoke the appearance of a color of specific hue and intensity. The trained eye may find that precise color among the 202 possibilities on a sheet of paper. Without conscious thought, the eye can be logical, when its capacity to respond to boundary changes induces structured sequences in hue and intensity. Intermediaries between colors as disparate as an orange and blue-violet can be obtained, by juxtaposing the two hues as flat papers, then visually selecting a logical intermediary. With this desaturated color placed between the orange and violet, the two strong hues will activate both boundaries of the area between them. The color perceived in the middle field, against the first, will suggest the choice for the next mixed color in the sequence, by induction of contrast at the boundary. In this way, progressive sequences can be structured with any number of combinations of flat colors appearing as transparencies.

3-14

3-15

Once a perceptually logical sequence of four had been selected, and observed on a two-dimensional field a square was cut from the center of each plane, sequentially diminishing in size. *Figure 4a.* Placed in the space-frame a series of colored masks is seen one behind the other in space. If the sequence gradually diminishes, these four planes, separated in depth appear as a longitudinal space. If the sequence in size is irregular, each plane may appear to contain a small square within it, the whole appearing flattened, rather than spatially separated in depth.

If the brightest color in the sequence is placed against the rear wall of the frame, it tends to move through the sequence to the level of the window. When the squares diminish in area by regular intervals, the sequence can appear to turn inside out. *Figure 4b.*

The bright color is perceived in this case as the smallest of the squares, surrounded progressively by frames of the next three colors. In such a sequence a gray or desaturated colored window may appear ambiguous; located closest to the eye, it can appear to recede.

There is a special effect obtained by the conversion of surface to film. When the eye is sensitized, the film experience is readily induced. A bright color, per-

ceived as the smallest in the center of a field of progressively less bright colors, can appear in time to radiate spatially. When this occurs, the entire group of colors is influenced, in this case greatly enhanced in brightness. When a less intense color replaces the bright one in the center, then the entire group appears in turn, less luminous.

The contrast effect then, is highly interactive in the perceptual groupings of four colors in space. In a sequence where all four colors are relatively equivalent in brightness, the fourth color, appearing in the middle of the field will be influenced to appear darker or duller. In a sequence of saturated reds through oranges, the last orange in the sequence appeared brown. The contrast in this case worked so that the three nearer planes, which were bright, became a context causing the farthest to appear darker. The quantity of brightness in the field colors in this case influenced the contrast effect.

Less rigorous or logical color sequences give rise to a variety of spatial effects. A bright color placed behind one relatively less bright, will tend to advance to the position of the latter in space, but no further. Sequences of four colors without logic in their grouping appear to be spatially ambiguous in the window.

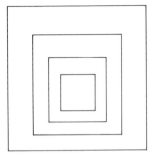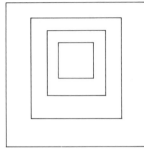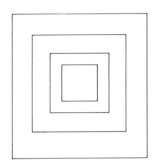

Fig. 5. Apparent spatial locations of four color sequence in window.

3-16

3-17

What is significant in this group of experiments is that the perceptual sequencing in hue and brightness between the colors matter, and not their actual positions as planes in depth. Providing that the perceptual intermediaries appear consistently as a logical series of steps, they can, in fact be located anywhere within the space-frame. The progression will work with equal effectiveness, and as logically, if it is reversed. The figure/ground or near/far locality of planes was determined by the contrast effect of film colors.

We found, however, that the illusion of spatial advancing or receding with perceptually mixed colors can be reinforced or diminished by their placement within the vertical/horizontal frame of reference. When seen as concentric groupings, with all intermediate areas appearing equal in size, the spatial effect is diminished. When aligned with unequal amounts of areas or bands of color visible in the upper and lower fields, the perspectival spatial placement reinforced the illusion. With a large area of the colors visible along the upper borders, the groupings seemed to recede in space. If this is reversed, and the lower

area increases, then the same sequence of colors seem to advance. *Figure 5.* These figural or formal changes are achieved also by moving the observer's vantage point; from a distance of 2 feet from the window frame, if the position of the eye is raised or lowered 3 inches, the relationshisp between the positions of the planes change their dynamics.

These readings occur only with perceptually logical groupings. When a color sequence perceptually opposes the perspectival illusion, the spatial effect becomes ambiguous.

Most compelling in these experiments is the intensity of the contrast effect. Colors appear enhanced in brightness when their surfaces are perceived as films. At the boundary the contrast effect is intensified by juxtaposition. When observed for a prolonged length of time the after-image is stimulated and the entire field appears brighter. This affects the color aligned in the interior space as well as the window frame, all hues appreciably increase in intensity, to appear nearly incandescent. These three-dimensional phenomena cannot be stimulated by photography.

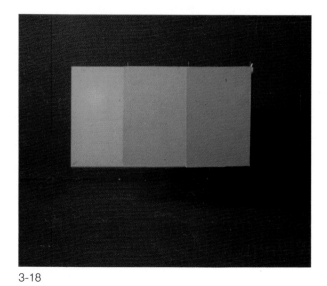

3-18

3-19

3-20

Experiments with sequences of hues and their intermediaries in space gave rise to free problems in the shaping of space with color. Illusions of truncated cones, tunnels, or pyramids were tried as extensions of the problem. In these solutions the relationships between hue and brightness contrast behave as colored light filling the intervals between planes of surface colors. The eye reads these as simulated volumes or interior spaces.

Most of the observations were made binocularly. Monocular observation tended to mitigate the spatial and depth effects, or to increase their ambiguity. The intensity of the contrast effect and influence of the after-image increases with prolonged observation. The ability to discriminate boundary conditions is an analytical one which develops with practice. While sensitized students can observe both boundary effects and the after-image more readily than untrained observers, experience increases responsiveness and sensitivity.

Studies using perceptual intermediaries were done in large scale and without the limited frame of reference of the window. A space-frame 4 feet wide by 4 feet high by 8 feet deep, was constructed and painted black. Three surfaces were left open; the top, front, and one side, to permit lateral or longitudinal placement. To test the influence of depth at contour, in this context, three planes were placed at 12-inch intervals relative to one another.[44] Selecting a group of three flat hues that perceptually relate as intermediaries, I arranged these as a sequence with the appearance of two overlapping planes. Working with the size/distance relationship, the plane closest to the eye was small in area, the next two were increased geometrically, so that the most distant was largest. At a critical reading distance, the three planes appeared as two transparent colors, overlapping in space, 3-18-19.

3-21   3-22

Color changes in space were also tested in the space-frame. Two planes, the same size and color, placed in a spatial context of two contrasting colored backgrounds, will change to appear as two different colors. If the two test areas are completely surrounded by the environmental field this will be the case. If they appear to overlap their respective fields however, the test colors will retain their identity and appear the same.

David Katz, who was interested in color constancy as a phenomenon, observed,

> The contrast effect on any one part of
> the configured field is not determined
> merely by the character, size, proximity
> of the other parts of the field; it is also
> of significance within which total config-
> uration it functions as a part.[45]

In the first experiment, the illusion of transparency was effective, despite the fact that the planes were spatially segregated in the depth of the field, once they could be perceived figurally as overlapping. If one shifted position relative to the space-frame however, the transparency effect vanished, and the

flatness of the colors were restored. Thus, transparency and three-dimensional space are effective only with relation to a configuration that contains or supports the color illusion.

Color Stereopsis: A Spatial Interaction

A compelling spatial illusion with color was discovered by experimenting with perceptual intermediaries. If three such related colors are placed against a two-dimensional plane, aligned so that the contrast effect of the first and third appear in the middle color, adjacent to their mutual boundaries, the eye and brain will read the illusion as that of two colors intersecting. The transparency of the intermediary color caused by boundary relationships with the two adjacencies, occurs when groups are related sequentially in hue and brightness. Transparency accounts also for the phenomenon of overlapping.

To test the three-into-two illusion in the space-frame, three perceptually related colors were placed as parallel planes sequentially in the field, one 3 inches from the window, the second at

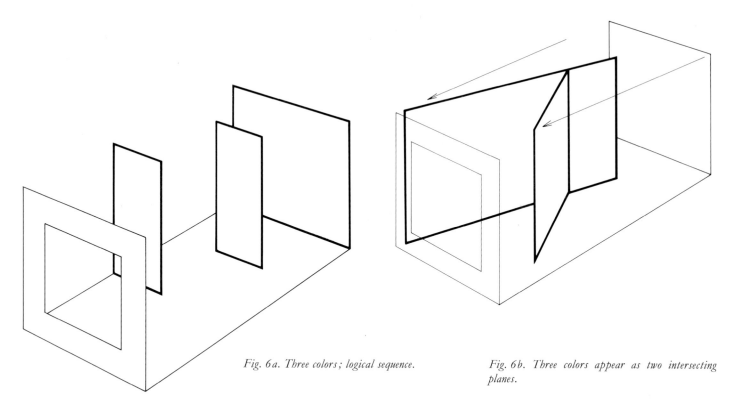

Fig. 6a. Three colors; logical sequence.

Fig. 6b. Three colors appear as two intersecting planes.

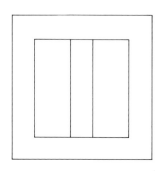

Fig. 6c. Appearance in window.

6 inches and the third 3 inches behind the aperture. *Figure 6a.* All three were observed through the window frame, where at first they appeared parallel to the window and adjacent to one another laterally. After a few seconds, the brightest of the group appeared to change position and move forward in the field, one edge in front of the window frame, while the second remained adjacent to the middle color. This plane in the sequence, an intermediary between the first and third in color, appeared increasingly filmlike. Now the third color appeared to change position. It moved forward in space diagonally, to appear slightly in front of the window opening, along one boundary, while remaining adjacent to the middle plane, on the other.

After 30 or 40 seconds of prolonged observation, the diagonal shift in the appearance of the two planes was enhanced by the transparency of the middle one. As it increasingly appeared as a mixture of the two adjacent to it spatially, it eventually disappeared. When this occurred, as a rather sudden and dramatic event, the total image fused or set as a new and unique visual pattern. Two colored planes were observed intersecting diagonally through the field. *Figure 6b.* The observer experienced the tension associated with stereoptical viewing, as fusion took place.

Many versions of this illusion have been tried, with different color groupings of three. These are always related sequentially, as logical intermediaries in

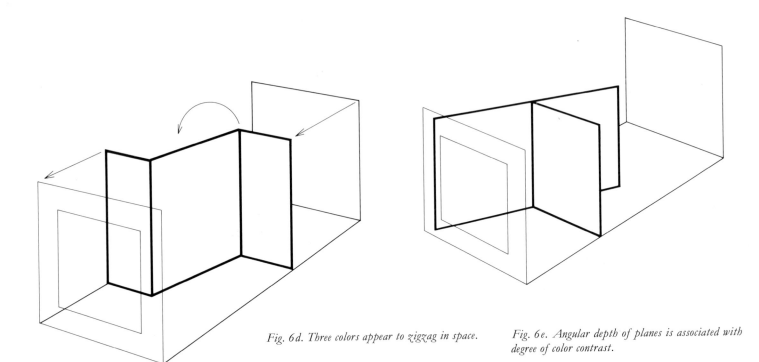

Fig. 6d. Three colors appear to zigzag in space.

Fig. 6e. Angular depth of planes is associated with degree of color contrast.

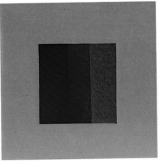

3-23

hue and brightness, although steps in distance or intensity can vary. Hues of similar brightness must share a boundary with a perceptual intermediary between them; if the middle color is brighter or darker, the transparency effect will not take place. Each logically related sequence obtains a illusion when three colors are perceived as two.

In some instances, when the three do not relate perceptually as mixtures of hues but are monochromatic intervals of brightness, the illusion results in a three-dimensional zigzag in space, with the middle color only shifting to the diagonal position. *Figure 6d.* Once the effect is perceived, the observer can shift his head to the left or right, and the illusion persists. It then appears to rotate, as diagonals do in perspectival field.

Although it takes a radical reordering of expectations to perceive it, the effect has become well established in experience. The diagonal directions are observed to be sharply defined or more shallow relative to the window, depending upon the boundary conditions between the three colors. *Figure 6e.* Distant or close intervals in hue and brightness are the gradients that account for spatial discrepancies.

The diagonal intersection appears after prolonged, binocular observation. One experiences a tension similar to that attained by looking at a stereopticon. But whereas that classic illusion is the result of the fusion of the two separate images that fall on the retina of each eye, this effect is obtained by the direct, simultaneous stimulation of both. Here the spatial reorientation of the planes occurs when the three colors appear to fuse as two.

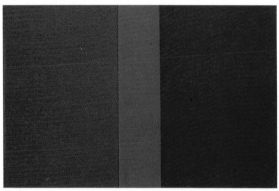

3-24

Throughout this series of experiments it is evident that there is a significant relationship between the perception of the boundaries between adjacent hues and the visual experience of space. The location of film colors is always perceived relative to a boundary. In the case of the spatial intersection, the formal simplification or resolution of three colors as two by the visual system occurs as a reorganization of the spatial direction of the colored planes. In previous experiments, the reorganizing has occurred in their sequence or progression, which as we have seen, can be a contradiction between true and perceived location. We have come to regard color then, as an important aspect of spatial perception, since in their juxtapositions colors can result in bounded fields or planes which can be interpreted to have direction, orientation, and location, the properties of visual space.

Complementary Hues and the Visual Deletion of Boundaries

Testing the effect of three colors, in which the intermediate mixture represents the additive result of two different surrounding hues, the following problem was posed:[46] What happens if two complementary colors are set on a plane, 2 or 3 inches apart? If the observer stares at them prolongedly, with both eyes, then the negative after-image of each is activated and they begin to mix with their respective hues. For example, after a red-violet (R V R t-1 of the Color-aid system) was placed on the left and a yellow-green (G Y G t-1) on the right, a lighter middle gray (Gr. 3) was selected from the color pack. All three were pasted adjacent to one another on a board with the gray in between the two hues.

From a distance of about 2 feet, the negative after-images of the two saturated hues were activated and began to mix and to overlap their respective stimulus areas. Since in this case each hue represented the other in its negative after-image, when mixed with it, each appeared in its respective areas as the same lighter gray. Once the intensity of that gray had been observed with prolonged observation, a lighter gray, which appeared to be equivalent to that which had been experienced as the consequence of the mixture of hue and after-image, was selected to replace the gray between them. After about 2 min-

utes, the boundaries between all three colors began to fuse, the entire area of the field appearing to be the same light, almost whitish gray. The eye will not remain fixed; prolonged attention causes involuntary muscular reactions, so as the eye moved, the boundaries between the colored areas would appear and disappear. But after the eye was fully responsive to the negative effect, and its mixture with the stimulus, the appearance of a uniform field was observed, with flashes of boundary occurring, after prolonged exposure.

In this experiment, the elimination of boundary is the consequence of two complementaries surrounding an additive mixture. It is a more difficult problem to resolve than the three-dimensional diagonal intersection, since the eye seems to seek a boundary, with whatever consequence to the perceptual organization of the pattern there may be. This elimination or cancellation of boundary also represents a reduction or simplification of the stimulus. Thus a complex illusion can result in a simpler organized pattern to the eye.

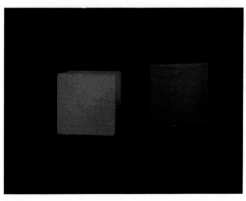

3-26

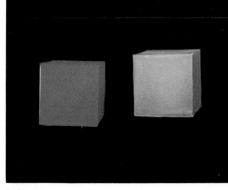

3-27

3-25

Space-Frame Experiments

The relationship of color to attributes of size, placement, and distance was studied in space. In the frame previously described (a construction 4 by 4 by 8 feet in dimension, painted black, and open at three surfaces, at a lateral and longitudinal end and at the top), cubes and planes of different sizes were studied with reference to its parameters. Varied in color, but uniformly illuminated, the series of experiments were structured and observed, for the most part, monocularly.

This group of studies was based on the premise that the psychological processes underlying the size, shape, placement, and color of objects, are not separate and distinct functions, but are interdependent parts of the same general process of perception.[47]

Experiment 1: Red/Blue

Two cubes, of unequal size and equal brightness, one blue and the other red, were placed at a distance from each other to appear the same retinal size. 3.25. While they appear aligned in a horizontal plane, there is a strong tendency for the red to advance, the blue to recede. In 3.26 the blue cube has been moved higher in the field, resulting in a stronger illusion of spatial recession.

In fact, the red is larger (9 inches) than the blue (7 inch) cube, and is placed in each situation described, farther from the eye than the blue. Though physically closer, the blue appears to be farther away in each case.

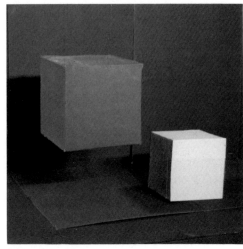

3-29

In a red background, two figures of different size, a large red and small yellow, are placed in a space-frame. In its context the yellow cube segregates from the red field and appears more pronounced than the red cube. When the yellow object is apparently related to the far edge of the frame, its location becomes more ambiguous. The illustration shows the spatial position of the yellow more distinctly, because it is related to the horizontal boundary of the surrounding field, near the eye. In all cases however, the red cube was physically separated from its background, and closer than the yellow to the observer.

3-28

Experiment 2: Figure/Ground

Contrasting colors in the environment influence the appearance of two red cubes of equal size. Though physically nearer, and consequently larger in the visual field, the red in a red surround tends to dissolve into the background and appear smaller, by comparison with the cube seen contrasted with the black field.[48]

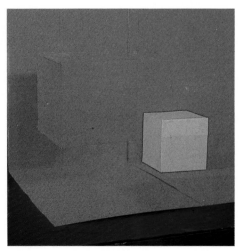

3-30

3-32

3-31

Experiment 3: Size and Placement

The influence of the interaction of color and size on apparent placement is studied. In 3.31 a yellow cube has been placed behind, but higher, than a smaller blue object in the space-frame. Its greater size and intensity causes the yellow to appear closer. When the yellow and blue cubes appear similar in size, placement in the field determines the spatial effect. 3.32. The yellow, lower in reference to the edge of the frame, appears to be closer. In both cases the blue cubes were closer physically to the eye, but the cues of apparent size and relation to the field appeared more significant.

Experiment 4: Size/Color

In this experiment, a small green cube is placed lower in the field than one colored red. The proximity of the red cube to the eye belies its true location; it appears to shift position within the field, from the floor to a plane suspended above the green.

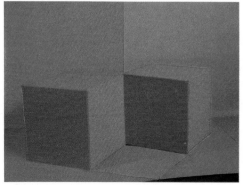

3-33

3-34

Experiment 5: Volume/Void

The field is split, one half green and the other red. Two red cubes are placed, so that one is contrasted with the field, the other integrated with the background. A green square, placed between the two cubes, causes a third, convex cube to appear. The boundaries of the red cubes, conditioned by their juxtaposition with the background colors, cause the shape of the red in the red field to appear softened, and less distinct. By comparison, the cube seen against the green field appears more solid. After prolonged observation, however, the vibrating boundary between the field and cube affects its solidity. As these edges become more active, neither cube seems as visually convincing as the "fake," between them.

3-35

3-36

3-37

Aspen: A Large-Scale Illusion

An experiment done at the University of California at Los Angeles provided an opportunity to test the intensity of hues in the field. A group of graduate students designed and constructed a large-scale illusion. Playing on a word, *Aspen*, its elements formed a geometric projection. The spatial field was 40 feet wide by 100 feet deep, the five letters were three-dimensional composites of forms displaced in the field. Four hues were used for the word, and they were juxtaposed against a fifth, uniformly colored background. In the ambient light of the environment, the four hues appeared to be fully saturated.

The assemblage was oriented so that all units faced the same parallel direction. The five letters were dispersed in depth as three-dimensional objects. *Aspen* was visible clearly only from one position, a hole in a screen, representing the single vantage point; elsewhere, the pattern appeared fragmentary. Each letter was split, to cause further ambiguity and to disperse the colors of the letters

3-38

3-39

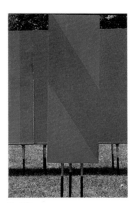

3-40

3-41

throughout the depth of the field. For example, the letter A was built as a three-faced rectangular solid; against one face a blue triangle was painted. From a distance of 80 feet a second, smaller blue triangle appeared to fit within its boundaries, to form the negative space and horizontal bar of the letter A.

The E, designed as a box, had against its open face, three red, horizontal elements. When seen juxtaposed with a plane, located 10 feet distant in the field, a cadmium yellow, in shadow, at the interior of the box, appeared equivalent in brightness to the yellow ochre background. When juxtaposed, the

boundaries between box and field were deleted, and they appeared uniform.

To elaborate the disguise, the letters were playfully manipulated. The identity of the A was referenced on a face of the solid visible from the rear of the assemblage, by a large, elaborate, calligraphic A from the Book of Kells. The S and the P, the only two letters to share a plane, were combined on the panel farthest from the eye, like a snake, 8 feet tall and 16 feet long, the head formed by the P, and the tail by the S. An intersecting vertical element changed its meaning, when it fell into position from the frontal view, and separated the S from the P, by dividing the head from the tail.

The artful illusion played perceptual tricks. On the whole, it worked upon the size/distance relationship. While the three-dimensional elements ranged in the vertical from 8 inches to 8 feet, they appeared equal in height from the central vantage point, when the word *Aspen* appeared in a single, horizontal plane.

In addition we tested the persistence of color intensity, namely, that of saturated blue, cadmium red, red-violet, and yellow ochre. As these had been dispersed at least twice in the visual field, comparisons between them had been

3-42

made within varying distances within the 100 foot frame of reference. Hues were separated by intervals of between 20 and 80 feet from the eye. Within these dimensions, the hues described maintained their saturation.

The brighter, more saturated yellow might have been expected to persist in its intensity, but it was less certain whether the unstable red-violet, and desaturated yellow ochre would change at a distance. The contrast effects between letters and their background were sustained, despite their displacement in space, as long as the total pattern remained coherent.

The assemblage remained in situ for the period of two weeks, providing the opportunity to observe it at different times of day, under changing angles of incidence, and intensity of sunlight, and under an overcast sky. As long as the total context changed (from bright to subdued daylight), the uniform appearance of the pattern remained the same. Under certain circumstances, however, hues were independently affected by the change in luminosity. At twilight, the blue appeared to intensify, becoming uniformly lighter than it had appeared at noon. The Purkinje effect, a phenomenon occurring at the threshold level, explains the change, when the eye responds to the reduction in ambient light intensity by experiencing blues and blue-violet as brighter hues.

The color constancy observed in the Aspen illusion may have been significantly influenced by the environment. To test the influence of the regional context, the assemblage should be reconstructed under less luminous conditions of ambient light, in the Northeast. As the assemblage was rationally placed with relation to the sun, its position supported color consistency. Frontal and parallel to the sun, as the light source changed, so did the appearance of the colors, but in a regular way. If color is integrated within the spatial coordinates of the field, it can be a unifying factor in environmental or architectural design.

Ralph Knowles has studied light and developed highly sophisticated ecological surfaces as a response to natural forces.[49] In his work, architectonic form evolves as a sensitive reaction to the light of the environment, resulting in complex, energy-conserving structures.

Controlled color studies on an urban scale should be attempted, in different regions. Reference was made previously to the occurrence of urban color with distinctive characteristics in places where the eye has reacted to ambient light. In turn, the use of color in the urban environment has significantly affected mood and defined its character.

3-43

The factor of time as an influence on perception is a significant but little studied phenomenon. The environment is not static, but a dynamic continuum, experienced serially, or disjunctively. Color can be significant as an articulator of time, because of its paradoxical nature. Since hues differentiate and localize, they arrest the eye or articulate space; they can sustain their identity in time. The spatial continuum of a street has been varied and identified by the repetition or frequency of hues; colors can carry from one spatial field to the next, by overlapping or recurrence. Color is a very significant aspect of pattern, as we will show experimentally, in three as well as two dimensions. Aspects of pattern-making in perception can be extended to temporal frames, a prospective enrichment of the urban place.

Color can shape or alter space; accordingly, its dimensions can become a sensory basis for building form. The designer, as well as the painter, can shape the urban or environmental field by considering color at the onset of the design process, integrally, as one of its constituents.

The relationship between the arts of painting and architecture were sought by the De Stijl painters in the Netherlands. They recognized the significance of color, and attempted to integrate it with architecture. Mondrian's treatise, *Plastic Art and Pure Plastic Art*,[50] presents a utopian view of the unity of the visual arts. Formulating his relationships with "pure" primaries, Mondrian's painting clarified and reduced the psychological framework to its essentials. Vertical and horizontal composition related to the experience of gravity, and the "reality" of forces underlying visual experience. A reduced palette enforced the plasticity of the primary hues, in their propensity to express and clarify spatial relationships. Mondrian's design for a room identified the universal coordinates of dimension, with the sensory limitations of the psychological primaries, red, yellow, blue, black, and white.

3-44

While a reductive palette is characteristic in the work of these Dutch artists, its usage arises from a context that is entirely man-made.[51] The use of primary colors in this case, then, may represent a significant environmental response

But a more direct, sensuous response is evident in the work of the Mexican architect, Barragan. He includes colors so integrally that they appear to be an intrinsic part of form. Redness becomes wall. In the predominantly natural environment of Mexico, where pervasive sunlight and vast spaces influence the eye, his architectural forms seem to have evolved simultaneously with their color and placement.

3-45

The use of color in architecture now, phenomenal and global in scope, merits more than casual mention, but critical analysis. In the United States, Michael Graves reflects a response to the region of the Southwest, as well as to allusive historical references, in his color usage. Charles Moore combines color and materials with significant rhetorical elements. The contributions of the color designer, Tina Beebe, enhancing the spatial and formal attributes of Moore's buildings, are intrinsic, and not superficial. The works of individual architects, resembling the aspiration of painters and sculptors, give rise to personal, metaphorical statements, in which color is again significant.

Painting and architecture were closely associated also in the work of Theo van Doesburg. His palette extended to include secondary colors, purple, green, and orange, in seeking the integration of color in architecture. While much of his work remained theoretical or is extant in drawings, Van Doesburg regarded visual structure to be an inclusive process, determined by sensory, as well as measured dimension.

4   The Geometry of Brightness and the Perception of Form

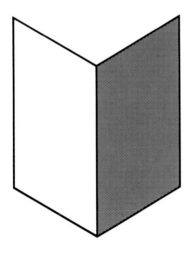

Fig. 7. 90 degree angle illuminated.

In the sensory world the perception of three-dimensional form depends upon the interplay of light and shadow. An angle forms precise visual patterns in ambient light, and these combined with their projected shapes can account for their reading as 90 degrees, 45 degrees, and so forth.

These patterns are so familiar that we tend to accept them uncritically, unaware that they represent *ratios* of lightness and darkness. From experience we can see the difference between an acute and an obtuse angle, and estimate their depths. This is, in fact, one of the essential functions of the eye and brain, or what I have come to call a process of quantification without number.

To test the effect that light and shadow have on the reading of angles in space, a right angle is constructed of white mat board, each plane 6 × 9 inches in dimension. When it was set on end, and illuminated it from one side with a Luxo lamp, the differences in contrast obtained by moving the lamp at fixed distances from its surfaces, were observed at distances varying between 6 inches and 6 feet. *Figure 7.*

Knowing that the intensity of light varies in an inverse proportion to the distance of its source, we expected to observe an increased contrast between the illuminated and shadowed side as the lamp was moved closer to the surface. Setting a light perpendicular to a plane at a distance of 3 feet, the observer stood 10 feet from the model. From a position midway between the two planes the 90 degree angle was observed to have a light and dark face of the same size and projected shape. Time was allowed for visual adaption to take place, then an estimate was made of the difference in light and shadow between the two faces. The contrast observed was simulated by using the white Color-aid paper and selecting a gray from its scale of 18, placed adjacent to it. From each paper a shape, a two-dimensional analogue of the angularly projected face of the model, was cut and pasted to the board. The two-dimensional rendition then, was compared with reality. When a contrast between the two shapes had produced a boundary visually equivalent to the one observed as the angle between the illuminated face of the model and the one in shadow, we had produced a convincing chiaro e scuro pattern.

Had the choice of papers produced a contrast that was exact? By placing them against the model itself in reverse order, maintaining the same incident light, would the gray paper resemble the white surface of the unilluminated plane? What is the proportion of light and dark in the contrast perceived on the model? Is there a relationship between this contrast and its simulation with a scale of values?

Most students thought that a test of the analogue against the model would be inaccurate, a few thought that it might match the shadow, but none could explain why. Therefore we tested it. The gray paper chosen to represent the shadow in the chiaro e scuro rendition was placed against the illuminated surface of the model. The discrepancy between it and the value of the shadow was so great that, without altering the appearance of its acuity or the degree of its slope, the 90 degree angle appeared now to be a gray, rather than white model. Almost immediately, a contrast boundary appeared at the juncture of the angle, with the gray paper placed against its illuminated face. It was this boundary condition that we sought to change; the criterion for observing contrast between shadow and gray. To soften the appearance of this juncture, darker grays were then selected from the paper scale, and replaced against the illuminated angle. Since we were aware that the observed effect of contrast is at variance with measured reflectivity we continued to try to achieve the match visually.

Through trial and error darker grays were tested and compared at the contrast boundary and with the shadow on the opposite face of the angle. The value of the original estimate, by comparison with the descending order of grays was too light by half. When a gray was found which matched the shadow, it was about twice as dark as the one originally chosen to represent the contrast in the two-dimensional analogue.

When placed against the plane the near black gray effectively cancelled the contrast; there appeared to be little or no boundary on the 90 degree model. The only clue to angularity remaining was its projected shape seen against the background of the table. The result of the new visual construct was disturbing. When the two sides of the angle appeared equal in value it could no longer be seen as a three-dimensional form but appeared, instead, ambiguous. A tension arose between the angular clue to form and the surface values. Viewed monocularly the tension increased, as the angular projection was read as a shape floating against the table top. When a mask was cut as a rectangular hole in a white paper, through which the model was observed, there was no longer any ambiguity in the comparison.

With its edges concealed, the two faces of the 90 degree angle now appeared as two equal grays. They appeared frontal and parallel in the frame of the mask. While one knew that the boundary between the two planes, scarcely visible as the threshold between two equivalent grays, represented the acuity of a 90 degree angle, this did not change the perception of flatness or of equivalence in value.

The contrast between the light and shadow of a 90 degree angle, then, is greatly underestimated by the eye. The degree of difference appears to be half, or a geometric reduction of the stimulus. With a gray scale of values the equivalent appearance of a white in shadow is obtained when the gray is about twice as dark as the shadow observed. This degree of contrast then cancels the distinction between the shadow and the dark gray, and the 90 degree angle appears as a flat, subdivided plane.

Value difference is only part of the color-form interaction. In the Introduction it was observed that there is no visible form in nature without color. How does hue interact with the geometries of brightness?

Leonardo da Vinci had commented on this in his notebooks

> As regards the equal diffusion of light, there will be the same proportion between degrees of obscurity of the shadows produced, as there is between the degrees of obscurity of the colors to which these shadows are joined.[52]

Differences between fully saturated hues are visually expressed in the Munsell system by their radial location relative to the scale of brightness. To test Leonardo's assumption, we selected a fully saturated bright yellow paper, and applied it to both sides of the white model, with the angle of light incident to its surface as before. Its brightness was enhanced on the angular face adjacent to the light source. On the oppo-

site face in shadow the yellow was darkened and desaturated. By visually estimating the apparent contrast of a yellow paper, a yellow ochre was selected. According to the degree anticipated by the achromatic study with grays, we sought an ochre approximately twice as dark as the one observed on the model. When placed against the illuminated surface, it appeared to approximate the yellow in shadow. The match was improved in the process of trial and error, until the right yellow ochre was found equivalent to the appearance of desaturated yellow.

Testing red under the same light condition as the yellow study, we found two that matched in the chiaro e scuro pattern, to be light desaturatedred, and dark red-violet. A hue in shadow appears to contain its complement, and the hue is influenced in appearance by mixture with the light source. Incandescent light in this case added yellow to the red-violet, so that directly opposite the light, the red, in shadow, appeared more violet. A match was approximated first logically, taking into account these discrepancies in appearance. Dark red-violet placed against the illuminated face of the 90 degree angle appeared equivalent to light red in shadow.

Color variables, its hue, saturation, and brightness (value) are highly interactive visually. Fully saturated red, a color intrinsically darker than saturated yellow, appears less influenced against the angular model by shadow. There is no system which shows relativity in mixture accurately; Munsell's is the best approximation. Empirical testing with individual hues however, sensitizes the eye to notice subtle distinctions.

4-1

4-2

To aid visual analysis, the relationship between the variables was observed abstractly, by the visual equivalent of an algebraic equation. Using the white and gray papers which had caused the illusion in the first experiment, these two were placed adjacent to one another against a flat plane. Below them the reds were placed, the lighter under the white paper, the darker beneath the gray. The contrast between the two achromatic values observed produces a distinctive boundary. The pair of reds are observed also to have a boundary, as they contrast with the pair of values. When this boundary appeared equivalent, then that separating the two areas of red was accurate, or the distance between two reds contrasted to the same extent as those produced solely by value change.

When the set of reds was placed against the angular model, with its values in contradiction to the light-shadow pattern, the appearance of the model was modified. If the two adjacent surfaces of red appeared equal, a tension between the flattening surfaces and angular edges was experienced, as it had been in the gray study, and the model appeared ambiguous dimensionally. When these two colors were observed

through the aperture, with the angular boundary masked, two equally colored planes appeared in the same frontal position in space.

The study of a simple 90 degree angle yielded complex phenomena. Value is the fundamental building block of form. When color is related to the visual proportion of light and shadow, it influences the appearance of boundaries which constitute the experience of volume. Manipulation of these variables can produce controlled illusions or altered effects of the appearance of forms in light.

Discrimination between values, or visual estimation of the relative brightness of hues is extremely difficult, as Josef Albers knew. In a problem he called "reverse ground," the interaction of hue and value in colors was tried in a two-dimensional study. In this experiment three colors can have the appearance of two. Their relationship is specific; two colors are distant from one another in hue and brightness (value), the third is an intermediary, although not necessarily the subtractive mixture of the two components. In the example illustrated, a gray, midway in value between and a yellow-green and lighter

4-3

red-violet are selected. Two narrow strips of gray are placed in large fields of the two hues, placed adjacent to one another. After visual adaption, the same gray is influenced by its surrounding field to appear to contain the hue of its opposite field. What accounts for this paradox? The two hues are complementary as a pair, and the gray is desaturated, a weak stimulus. Visual observation of the study causes the negative afterimage of each surrounding field to become active. As the area of the gray strip is influenced by this optical change, it increasingly appears to contain the negative hue. Against the yellow-green the gray appears to contain red-violet, while the same gray against the violet field looks yellow-green. The precise match is a function of visual logic.[53] It appears also to be a geometric reduction, at least in the brightness (value) component. The gray is precisely midway in value between the two colored areas, by visual estimate, while in this example the two hue components in each color represent complementary pairs. In reductively assimilating the group, the eye and brain seem to prefer the experience of two, rather than three colors.

In form-building with light and shadow the visual system seems to be function-

ing logarithmically. Our analysis with color and the 90 degree angle appears to support a similar situation with respect to complementary pairs of hues. When an intermediary in brightness was found, two complementary hues appeared equivalent to one another, against the model. The illusion of reverse ground is attained in two dimensions when the relationship between complementary hues is supported by a brightness gradient precisely between them. Proportion is fundamental then, to form-building with color, geometrical with respect to the parameter of light, or the brightness of color.

An understanding of the juncture of volumetric planes, in this case a 90 degree angle, is an essential step to the articulation of volumes with color. Having developed a perceptual building block with angle, we proceeded to experiment with simple volumes, and complexes or assemblages in space.

Constellation of Cubes;
90 Degree Angles Repeated

An experiment was done to understand the relationship between simple volumes and color. The three-dimensional model consisted of twelve square planes. Constructed of white mat board, the planes were assembled as three modules connected at 90 degree angles to one another.

They appeared as three half-cubes, stacked, two below one above, in a pyramid. To play upon the ambiguity inherent in this assemblage, the three remaining planes were joined to the cubes, so that one of the squares placed horizontally appeared to connect the two half-cubes at their base, while the second and third squares were connected at 90 degree angles to the remaining half-cube stacked above them, adjacent to its left and right face.

Thus, the total assemblage played upon the potential alternating concave/convex ambiguity that results when all twelve of the planes are visualized from a vantage point where they look approximately the same size and shape.

The model was then illuminated at 45 degrees from one side with the flexible lamp, until a distinct pattern could be observed for all twelve faces. Now a consistent pattern appeared on each of the half-cubes. They were independently articulated by a light, middle, and dark value, but the degrees of contrast differed, relative to their distance from the light. By applying the analysis of the 90 degree angle effect to the cubes, with the scale of gray Color-aid papers (eighteen steps), an illusion was created. Precise grays were set against each face, in reverse order; that is the darkest of a triad of grays was applied to the lightest (most illuminated) face, while the lightest was placed against the plane most in shadow, with each matching its intermediary value. When all the grays matched exactly, they effectively cancelled volume. In this case, the entire assemblage tended to appear flattened.

Having established the pattern of values as a substitute for the pattern of light and shadow, we shifted the lamp to the opposite side. When fixed in the 45 degree position symmetrically opposite to its original location, the grays all appeared in reverse order, exaggerated in their geometric steps by the light added to them. Now the assemblage appeared to be frozen in light, reiterating its original angle of incidence. In addition, the concave/convex relationship was played upon. The perceptual contrast was clarified; that the appearance of volumes is proportional to the angle and distance of the light source was made graphically visible.

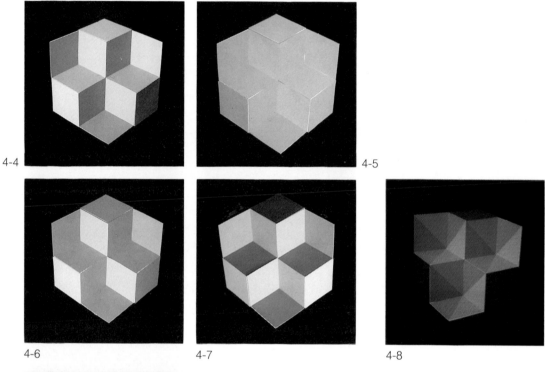

4-4

4-5

4-6

4-7

4-8

4-9

We altered the pattern next by using color; a red was chosen, with lighter and darker values, ranging from light red to dark reddish brown, to express the relationship previously established by the gray scale. Each face of the half-cubes was sectioned diagonally, and pairs of colors were placed against each of its faces to break their formal consistency with coherent new patterns.

In the last step of the experiment, each of the cubes was isolated by the application of a triad of values of a single hue. In this case the visual coherence of the structure is enhanced; the appearance of a threshold between hues articulates each cluster clearly, and an unambiguous pattern of convex volumes is seen.

Each step in the development of this series of experiments depended upon the proportion of light and shadow observed when the model was white. Once established, this pattern prevailed as the basis for alteration, enhancement, or clarification of form, when color was applied.

The Relativity of Form to Color

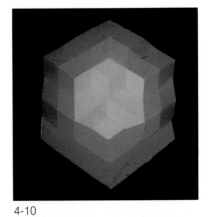

4-10

The components of this three-dimensional assemblage are belied by its appearance. Formed by three L-shaped solids of decreasing size, stacked one upon the other, they are set on a square plane; in their midst is placed a cube. The following colors appear on the faces of the configuration: orange-yellow hue, yellow-orange hue, orange hue, red-orange hue, desaturated orange (brown), and desaturated purple (brown-purple).

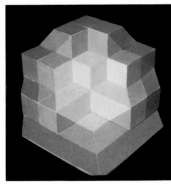

4-11

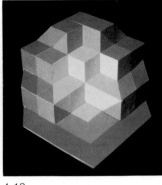

4-12

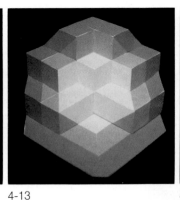

4-13

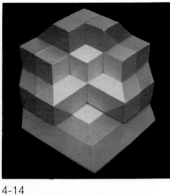

4-14

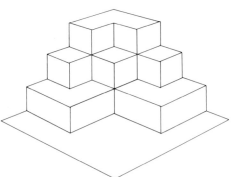

Fig. 8. Stacked constellation.

Constructed of white mat board, the small model is observed frontally, and from angles ranging over an arc of 180 degrees. Seen directly from a plane parallel to the eye, and illuminated from a 45 degree angle, the value pattern of light, middle, and dark was articulated.

Accordingly, a group of colors was chosen, related adjacently as hues, and placed concentrically on the model. The bright, fully saturated yellow-oranges and orange-yellows were placed in the center of the configuration, surrounded by red-oranges and browns, and at the perimeter of the assemblage purple-browns were placed in a descending order of values.

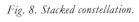

4-15

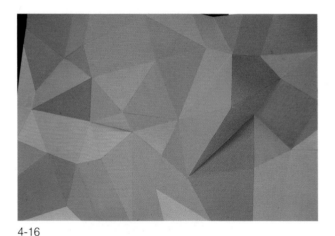

4-16

4-17

When these colors were placed as papers against the surfaces of the model in reverse order to the underlying pattern of light-shadow, they appeared nearly equivalent. As the boundaries of each ring-cluster of faces on the assemblage were resolved, it attained the appearance of flatness. When completely covered by Color-aid, rings of brown-purple surround rings of red-orange consecutively around a hexagon of yellow.

A change in the angle of light radically altered appearance. By shifting its incidence from side to front, the model appeared progressively altered, from that of a flattened surface, to one increasingly articulated by color. The integration of hue with chiaro e scuro pattern results in formal variations in three dimensions, here from flatness to crenellation.

Free Study in Form and Color

A series of distorted pyramids is constructed of mat board on a rectangular plane. The result is a three-dimensional relief of contiguous irregular triangles.

Colors were selected as clusters of contrasting hues, to "map" the surface; yellow, mixed with greens and ochres, contrast with sienna browns against blues and blue-greens, which in turn, contrast with violets, and violet-grays, desaturated by mixture. The surfaces were painted with colors rigorously related to the brightness gradients represented by the irregular relief, a series of concave and convex forms varying in height between 2 and 6 inches from the plane.

When the form of a surface is neither modular nor repeated, clues to form are very ambiguous. As a consequence, this model is radically altered in its appearance by light or by a shift in the observer's position.

4-18

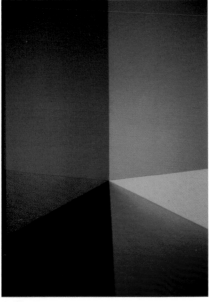
4-20

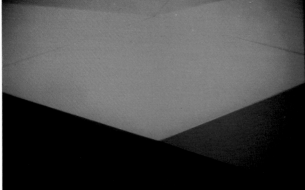
4-19

Interior Dimensions Perceived

The effect of color on an interior space was studied. A small room was constructed; its proportions based upon the module of a standard piece of masonite, 8 feet long by 4 feet wide. Using this module for the floor and ceiling, its horizontal planes were supported by two verticals at each end, and two pairs of panels formed the structure's longitudinal walls. An oblong interior space, 8 feet high by 8 feet long by 4 feet wide was created.

The small interior space could be entered, and when an observer sat on the center of its floor, the spatial coordinates were distorted by perspective. Observed first as a white space illuminated by a single incandescent bulb in the ceiling, the pattern of light and shadow was indistinct. These visual conditions suggested the invention of a strong composition using color, to modify the perceived dimensions of the constricted space. First the junctures of wall and ceiling were made similar in appearance, by painting a fully saturated red in the corners. The red continued along a wall surface, to bisect it diagonally and to appear to overlap the ceiling plane. Adjacent to the red, areas of gray and blue were painted in contrast. When completed the interior surfaces were redefined by color.

Diagonal lines separated the planes of walls, ceiling, and floor. When seated at its center the observer experienced a strong influence of the diagonal directions on the spatial coordinates of the interior. While visible, the vertical-horizontal planes were less emphatic, or distinct at their junctures, when compared with the sharp boundaries between the colors. Observers who remained in this position for 10 or 15 minutes, reported the experience of a radical reorganization of the interior. Its space appeared to be altered from that of a vertical horizontal, to a system based on diagonal coordinates.

The pattern of colors engendered in this case strongly influenced spatial perception. The monocular aspect of the photograph enhances the compelling experience by reducing it to abstraction.

5   Configuration, Pattern, and Dimension

How does color interact with form? The environment contains considerable examples and variations; visual transformation is common in nature and both purposeful and playful.

To the human eye the appearance of objects is determined by vantage point; without taking a position or stance, one could not assess objects in their relationship to one another in space. The visual field is rationally disposed when the eye judges the dimensions of objects relatively. Perspective reduces measured dimensions to their two-dimensional equivalents as projections; depth-space converts to field-plane. If color remains confined within the boundaries of figure/ground, it enhances dimension, but if an interplay of light, shadow, and color is contrived, then the visual estimations made between these clues to form may radically alter their appearance.

Visual transformation is the result. In a series of experiments called transformations, we played visual games with forms, by coloring their surfaces. Assuming the spatial position of the observer to be frontal to the object, and that the experiments would be observed monocularly, the following configuration devised was similar to and simpler than the constellation discussed in the previous chapter. Seven 6-inch squares were assembled to appear as two cubes, joined by a square plane at their common base. The squares were cut from white mat board; three were joined at a 90 degree angle to form a half-cube. Repeating this, a second group of three squares was formed and connected to the first half-cube. When observed frontally, the two forms appeared as two cubes joined at a 90 degree angle. The seventh square was then added to connect the two cubes along the lower edge between their inner left and right surfaces and at a 90 degree angle.

From a distance of about 10 feet, below eye level, the configuration appeared symmetrical. In the ambient light of the studio, a pattern of values, light, middle, and dark, appeared to articulate each plane, defining the cubic dimensions. With a directional light source, this pattern is a function of its angle of incidence. Except when the lamp is focused perpendicular to the frontal plane, any angle yields the familiar, ubiquitous three-faced pattern.

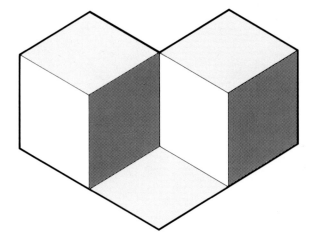

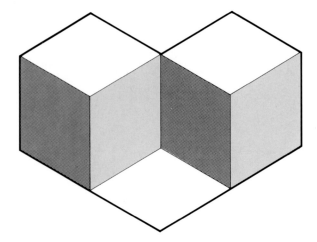

Fig. 9a. Seven-faced configuration in light and shadow.

Fig. 9b. Seven-faced configuration, changed angle of incidence.

In studio light and from a fixed, frontal position, two cubes appear to be connected by a square. Observed monocularly, after a few seconds, a middle cube appears, between the two. *Figure 9.* When this occurs, either two convex solids appear to be separated by a plane, or a concave cube appears in the center. The following fluctuation occurred after the third cube was observed. After prolonged, monocular observation, the middle cube seemed to lift from and reverse its position in space, and the positive/negative, or volume/void, experience was sustained. Once observed, the illusion persisted, even as the viewer moved in an arc 180 degrees in front of the object, and despite perspectival distortion.

The triad of values seems to be the most significant clue to form; once observed it can reverse sequence and remain coherent, even if this means the figure has to be turned upside down and inside out *Figure 9b*. Prolonged attention tends to enhance perception of the center[54] and to isolate the area from

its spatial context. This may account for the fluctuation. A radical alteration of the cube's position and implied redirection of the light source are, after all, its consequences.

Once seen, this ambiguous configuration can be played upon, its articulation a function of the position of color. The perception of forms in space depends upon the resolution of all visual clues, light, shadow, hue, and boundary, as a total pattern. If color can significantly influence boundary, then three-dimensionality can be contingent upon it.

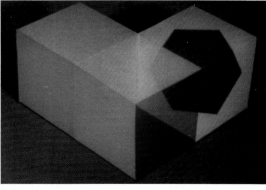

Fig. 10. Transformation 1: intersecting cubes, one transparent, the second flattened.

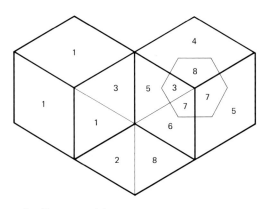

1 Orange-red hue
2 Red hue
3 Orange-red, medium light
4 Orange hue
5 Yellow-orange hue
6 Brown
7 Medium dark gray
8 Dark gray

The model was illuminated from two sides and above, causing the light/middle/dark pattern to occur top/side/side. The orange-red hue was placed on all three faces of the cube on the left. The observed change in light on its horizontal upper plane is real, not illusory.

On the cube on the right, two adjacent faces were covered with identical papers of yellow-orange. To compensate for the relatively higher intensity reflected by its upper surface, the third, horizontal square was covered by a fully saturated orange hue. Thus a precise match is obtained for all three faces.

A pattern was then imposed on the middle "cube". Since it shares a surface with each of the two half-cubes, their adjacent inner faces were modified. A triangle was cut from a paper of the light orange-red and pasted to the right face of the cube on the left. It shares a boundary with the upper, horizontal plane, and now matches it. Next, a triangle was cut in the red paper. It was placed on the floor of the assemblage, against the seventh square. Adjacent to the orange-red plane, at a 90 degree angle to it, the illuminated right surface

of the left cube and the warmer orange-red appear equivalent. This interplay of color has the effect of transforming the entire left of the assemblage to that of a rectangular solid, projected into space.

The grays are then cut into sections, which when perspectively projected from the corner of the cube, appear as a flat hexagon. These grays are placed in relation to the light source to appear equivalent, in accordance with the flattening pattern of the oranges.

At the intersection of the hexagonal orange shape to the right, and the projecting rectangular solid, two more colored areas complete the illusion. A brown, cut as a triangle, and placed against the left face of the right cube appears transparent. A light orange-red, cut as a triangle, placed adjacent to it on the same square plane, appears illuminated.

The total assemblage has been transformed through the use of flat, uniformly colored papers. Two cubes are no longer perceived. Instead, one looks more like a flattened orange hexagon, surrounding a smaller gray one. The second appears as a projecting rectangular solid, which, as it intersects the adjacent field, appears transparent.

The monocular effect is compelling at near distance. From more than 30 feet from the observer the illusion is obtained with two open eyes.

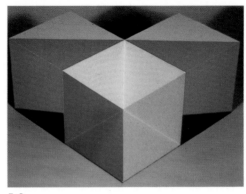

5-2

Fig. 11. Transformation 2: volume/void reversal.

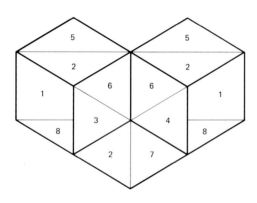

1   Red-orange hue
2   Red-orange, medium dark (brown)
3   Red-brown, medium
4   Yellow-orange hue
5   Orange hue
6   Yellow-orange, light
7   Yellow-orange medium
8   Black

This model has been illuminated similarly to transformation one, with two lamps, from left and right and from a 45 degree angle above, making its upper horizontals brightest. A red-orange hue is placed against two vertical faces, left of the left and right of the cube on the right. A darker red-brown covers the upper horizontal planes of each cube, its cooler and darker hue, reflecting a greater degree of light, appears equivalent to the warmer and lighter red-orange in shadow. Rigorously selected, these two colors are so close that there is no boundary visible to the eye at the 90 degree juncture. With the placement of these two hues on the four planes indicated, the solidity of the entire form begins to be affected.

Three orange triangles are cut, two are fitted on the upper horizontal planes of half-cubes, appearing to cut their symmetrical surfaces in half horizontally. The third orange triangle is placed on the seventh, connecting square plane, subdividing it vertically. As these three squares are all placed horizontally with respect to the angle of incidence of the light, they are equally illuminated and do not change appearance. The red-brown paper is cut as a triangle and

placed against the middle horizontal or seventh square, adjacent to the orange, to form a vertical line. From a light red-brown paper, a triangle is cut and placed against the right face of the left cube. As this lighter color in shadow appears equivalent to the illuminated red-brown paper adjacent to it, the 90 degree angle connecting the two surfaces vanishes.

To further "dissolve" the two adjacent half-cubes in the process of emphasizing the third, the left and right faces of this inner figure continue to be modified. Two triangles are cut from a light yellow-orange paper. These are placed against the equally illuminated adjacent inner faces, and remain therefore equivalent in appearance. Adjacent to the orange and red-brown, this color causes the illusion of a false top, added to the middle "cube," an apparently uniform field. Two small black triangles are added as final detail, to convert this field to a horizontal reddish-brown plane.

Viewing the model from the center, the observer sees a convex cube emerge from the middle, while a red-brown plane seems to run horizontally through the assemblage. When he or she shifts

Fig. 12. Transformation 3: two cubes become an L.

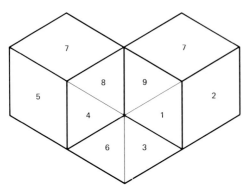

1 Yellow-green, light
2 Yellow-green, hue
3 Green hue
4 Green-yellow, hue
5 Green-yellow, light
6 Yellow hue
7 Blue-violet, hue
8 Blue-violet, light
9 Red-violet, light

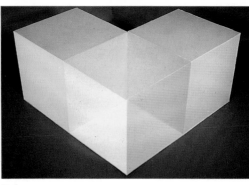

5-4

5-3

position, this cube appears to be crushed or distorted, while the reddish background plane apparently expands and contracts. Thus, the transformation achieved for a frontal position becomes a commentary on change itself with a new perspective.

The simplest of the solutions, that of an L-shaped solid, is obtained after careful analysis. The model is naturally illuminated. Oriented toward a north window in the studio, its faces are formed by a subdued light/middle/dark pattern.

In the three-dimensional original, the close adjacencies of colors, and their relatedness as near complementaries, imparted a filmlike quality, making this a particularly delicate solution. The nine hues, reduced to a pattern of three sets, are placed in this case with the intention of transforming the cubes into a single, regular figure. Therefore, the greens are distributed against the left and right sides of the cube on the right and appear also as a triangle, vertically intersecting the connecting seventh square. Adjacent to this, the three yellows are distributed in reverse order, relative to the intensity of the light, the

darkest against the most illuminated plane, and so on, against the cube on the left, and as a triangle completing the seventh square. The remaining group of violets are placed against the horizontal planes of the upper surfaces of the two half-cubes, and at their right and left faces respectively.

The most reductive of the group, this transformation is the result of a most rigorous search. The subtle steps in light/shadow pattern on the model, had to be matched with their equivalents in hue and light intensity in the choice of colored papers which oppose them. The transparency effect is quite complex, depending upon the evocation of contrasts of a particularly refined group of colors. The proximity of light intensities within this limited range of hues keeps their appearance closely adjacent. The yellows and yellow-greens are analogous in effect, while the light violets complement, by opposing the grouping. A delicate equilibrium between adjacencies and oppositions resulted in a solution of particular refinement, even though it is coherent only from one vantage point.

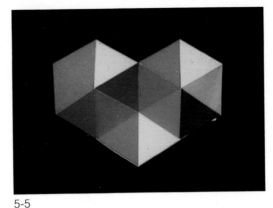

5-5

Fig. 13. Transformation 4: two cubes appear as three interlocking hexagons.

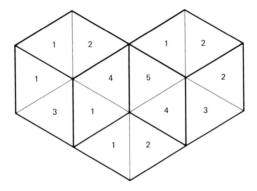

1 Orange-red hue
2 Yellow hue
3 Blue hue
4 Green hue
5 Violet hue

In this example, transformation is achieved by using fully saturated spectral hues, five in number. Each square plane of the assemblage is bisected diagonally, resulting in a consistent triangular pattern.

The colors are distributed so that the reds, occurring five times, appear in all but two faces, the yellows appear four times, the green and blue twice, and the violet once. The angle of incidence of the light is from above left; as a consequence, the red/green square on the

right surface of the cube on the left and the blue/yellow on the right face of the one on the right, appear darker and cooler. The colors are distributed selectively so that most of the adjacencies produce a hard boundary; an overall pattern of hues is the predominant effect, and the assemblage appears to be considerably flattened. In this case color boundary has replaced the boundary resulting from a 90 degree angle.

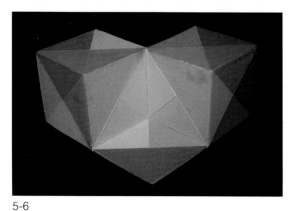

1 Red hue
2 Green-blue hue
3 Orange-yellow hue
4 Yellow-ochre light
5 Blue-violet medium light
6 Light gray
7 Medium gray
8 Medium dark gray
9 Red light
10 Blue-violet light
11 Green-blue light
12 Gray-violet light

5-6

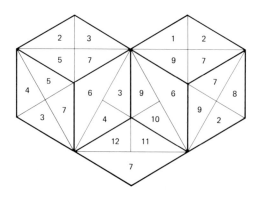

Fig. 14. Transformation 5: two cubes appear as three pyramids.

A complex configuration is the result of the play upon boundary conditions, in which color is distributed as pattern.

The model was illuminated frontally, in an attempt to minimize its apparent cubic dimensions. Four of the seven square faces are subdivided by diagonal lines forming four smaller triangles, three squares in the middle of the assemblage are subdivided into three triangles. The hues are distributed, so that each appears twice in the assemblage.

There were two solutions devised for the central "area." In one, a cluster of lightly tinted hues forms a pyramid in the center, and this is surrounded by a group of grays which appear to be equal. A transparency effect is engendered as the boundaries of the colored cluster are closely related to one another, and are contrasted in turn with the larger gray field. This effect is enhanced by the recurrence of the more fully saturated hues, distributed along the outer edges of the two half-cubes.

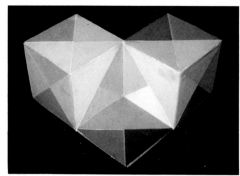

5-7

A more ambiguous and less consistent result is obtained by replacing the seventh square. In this example, a red/blue gray field displaces and disrupts the lightly hued pattern of a pyramid, with a more pronounced, saturated contrast. The new boundaries are harder; as a consequence the plane appears more broken.

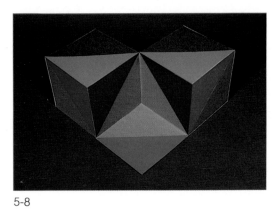

5-8

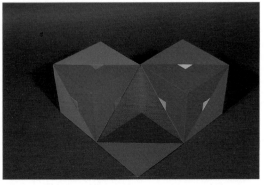

5-9

Fig. 15. Transformation 6: fluctuating volume/void.

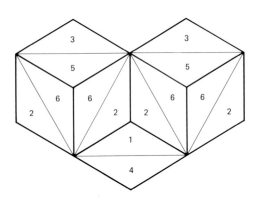

1 Yellow-ochre, light
2 Yellow ochre
3 Yellow-ochre, dark
4 Blue, medium light
5 Blue, light
6 Blue, hue

By diagonally bisecting each face of the assemblage, clusters of triangles cause the appearance of pyramids, and the selection of colors reinforces the dimensional effect. Here a pattern of light, middle, dark hues, grouped by yellow ochres and blues respectively, mimic the light/middle/dark ratio of illumination. Boundaries are maximized between the two complementary groups of hues, and their contrasting values. The consistent cluster of blues against yellow ochres sustains the figure/ground, or volume/void relationship. Two blue pyramids appear surrounded by ochre, or the illusion of a central ochre pyramid may alternate spatially with two surrounding blues. In this solution the assemblage is illuminated with two lights, from angles of incidence from above, and left and right of the model.

Except for the small triangular markers, the placement of diagonals in this solution is identical to those in transformation 6. Despite the fact that they are formally identical, their appearances differ as a result of color distribution.

In this case, the middle area has been emphasized by the juxtaposition of colors resulting in the hardest boundaries. The red hue, equally illuminated on the faces against which it appears, looks very pronounced, but remains formally ambiguous because it does not resolve into a single, simple configuration. The more coherent patterns of the clusters of blue-violets and blues articulate two pyramids, on the right and left half-

Fig. 16. Transformation 7: variant/fluctuating volume/void.

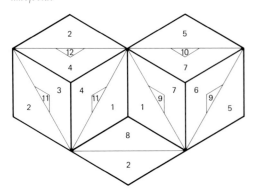

1   Red hue
2   Green-blue, light
3   Blue hue
4   Blue, medium
5   Red violet, light
6   Blue-violet hue
7   Blue-violet, medium
8   Blue, dark
9   Yellow, light
10 Yellow hue
11 Orange-red, light
12 Orange-red hue

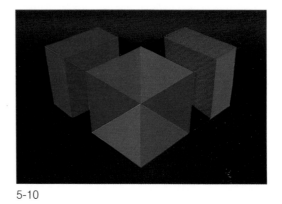

5-10

Fig. 17. Transformation 8: volume/void reversal.

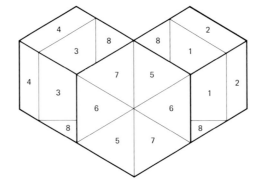

1   Red hue
2   Red light
3   Blue hue
4   Blue, medium
5   Red-violet, light
6   Blue-violet hue
7   Blue-violet, light
8   Black

cubes. In both cases the darkest value of these colors appears on the most illuminated face. Since two light sources symmetrically placed to the left and right of the model focus the illumination, the central reds of these faces look the same, but the red-violet adjacent to it on the right, and the blue on the left, appear as darker values.

The addition of the small and brightly colored triangles as ornamentation, activate the blue and violet fields against them. Thus the solution is rather tense and unresolved in appearance.

The center cube is articulated in this solution by the distribution of violet triangles on its three faces. By the use of two light sources, from left and right, focused on the center, the light, middle dark values of violet are distributed so that they appear as a hexagon. The reds are sequentially arranged, to appear flat in the center and dimension-

al on the periphery of the cube on the right. The same pattern is repeated on the left with blues. Insertion of four small black triangles complete the illusion that the two adjacent cubes are sliced vertically, making this a playful, if carefully contrived solution.

These examples of the transformation of the same three-dimensional assemblage give evidence that color is a very significant aspect of the perception of form. Although each of them is distinctive and unique, they share in common the change in surface patterns arising from the distribution of color, which in all cases supersede the volumetric base.

The changed conditions are particularly apparent at the boundaries between the colors used, which either contradict the underlying patterns of light and dark, or enhance them.

These changes can be characterized as follows; ambiguity arises in the perception of three-dimensional form, when a boundary caused by the juxtaposition of colors becomes more pronounced or visually significant, than those defining the structure.

In the series of transformations described, the same assemblage sets the condition for each illusion. Each solves dimensional appearance in a different way, by an interplay of attributes. A different visual intention was achieved in each case, by using a unique strategy. At times transformation was sought by contradicting the volumetric pattern, while others reinforced form by following, with color, an underlying pattern of values.

The basic configuration is transformed by a play on visual clues. In each in-stance the change represents a choice or visual intention. Just as forms can be changed, so can they be articulated, and by the same visual means. To create or to delete boundaries with color is the result of the proximity or relative distance between the colors juxtaposed and the underlying patterns of light and shadow. If these change, then so do the color patterns superimposed, but consistently, as we have shown.

Thus, while conceptually there is an opposite intention between *con*forming and *trans*forming, operationally the same thing is done, that is, to use color to build form.

Applications of the transformations just described may pertain directly to stage design, where a frontal position and control over the source of illumination are parameters. But architecture involves neither a static relationship between observer and object nor a constant light source. The experience of progressing through buildings takes time, and light fluctuates and is displaced in space. Enrichment of architectural surfaces however, can result from transformations of color and three-dimensional relief. An effect obtained by a static position or momentary condition becomes a series of images as the observer moves. The continuum of appearances, shifting with relation to the eye, is then a commentary on the experience of the architectural surface in time and motion.

6   Color and Form: Conform or Transform?

6-1

6-2

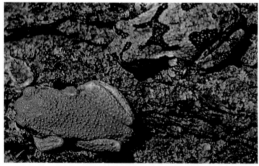

6-3

Ambiguity seems to be the result of highly visible and well-articulated visual patterns, arising from the distribution of surface color. Is this true in nature? Do patterns in nature conform to and define organisms, or do they confuse the eye by transforming essential volume beneath the surface? I think the answer is, they do both visual operations. In subtle ways and many instances, patterned color conforms and transforms simultaneously.

To begin with, boundary conditions in nature are the result of the interaction between organism and its environment. A moth on a tree, a fish in the sea, or a parrot in a rainforest have the same intrinsic figure/ground relationships with the natural world. The difference lies in their visual function in the environment. Against the tree bark the moth is concealed, because its marking is analogous to the surface texture of wood. Color contrasts make patterns on the animal which appear equivalent to those produced by light and shadow on three-dimensional structures of bark. These patterns seem to merge from a distance, superseding their dimensions and the overlapping edge between them.

Camouflage can work as concealment in another way. A brightly colored monarch butterfly, its highly contrasted orange and black coloration, can appear equivalent to a pattern of brilliantly illuminated grasses standing in an everglade. If the monarch moves to the sky, intense blueness casts it to conspicuous relief. Conceal or reveal, the visual game is entirely dependent upon how the organism interacts with the environment.

The principle of countershading is a strategy for the concealment of cylindrical forms in nature. In this case, light and dark patterns of coloration are distributed on curved surfaces, in a way to contradict the light source. An example of the inversion of chiaro e scuro, it has the effect, in ambient light, of arresting the animal by flattening its volume. Many small mammals have darker coloration on their upper coats than on their bellies, which are often white. Sunlight is directional from above so that the condition of cast shadow is pervasive below.

In the sea, countershading can work in the same way for opposite reasons. As the ocean depths become increasingly dark or murky, distant from the sun, brilliant colors may occur as bandings. The angelfish's brilliant fluorescent edges delineate it against a watery field. In ocean depths fantastic displays of luminescent color exist among species of fish that cannot see.

The elaborate patterning of colors playing upon the relatively simple shapes of fish enumerate species, defining them. The mocking strategy of an eye pattern on a tail is a form of concealment, as are speckled markings that mimic and record fluid patterns of the surrounding water. Scales are microstructures, developed to reflect or refract light, with relatively few cellular pigments.

The eyes of reptiles and birds are outlined or delineated by color, offering fantastic variations on the theme of reveal or conceal. The significance of the aperture, or the eye as a bounded field, is the visual commentary. Conversely, the reptilian eye appears to be cancelled or sliced by a colored line running through.

6-4

6-5

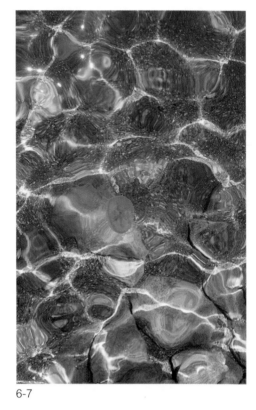

6-7

6-9

6-6

6-8

6-10

Pattern is so significant in nature[55], that its surface manifestation as the distribution and repetition of colored areas should be an independent study. In the individual organism it is genetically programmed, as much as is the skeleton or any other morphological feature[56]. Highly specific and clearly differentiated, they are found in enormous variety. Are there any basic rules or constraints to natural effects that might educate the designer? Again, does color conform to or transform the organism? Does the distribution of color work to make morphologic features, such as size and volume, more conspicuous, clearer, or more visible? Is this achieved by following the surface, its topography, or by opposing it? If color can enhance the character of surfaces, then is the perception of three-dimensionality contingent upon it? Does nature perform its permutations in altering its essential qualities by coloring them?

Autumnal changes of deciduous leaves, from green to the yellows, ochres, bronzes, Indian reds, cadmiums, siennas, and browns of October, are a dramatic but familiar example of natural transformation. In the process of dying nature seems to orchestrate a chromatic

6-11

6-12

coda. A closer look reveals significant effects or constraints. When a leaf changes color the photochemical effect is visible first in the cellular structure nearest veins and stems, then edges. The nutritional channels are contrasted, just as these structures persist throughout the life cycle as skeletons, once the moist tissue has been eliminated. The leaf thus tells its condition of visual transformation, as it conforms to a natural process.

6-13

At the same time, the integration of light and color is synonymous with the quality of place. The subtle grays of Manhattan, muted by the filtered light of its urban canyons, scarcely emit hue. By contrast, the canyons of the Southwest, layering geologic remains by traces of reds, ochres, browns, and buffs, reveal their time and evolution to the trained eye.

In the desert, where hue is bleached by radiant reflectivity, the conditions of place can be read. Dry beiges of soil or scrub are disrupted in spring by a lush flowering of magenta blossoms. Life ceases altogether, a symptom made visible in the deadly gray-green of erosion. Signs, symptoms, and conditions abide in natural color.

6-14

To the eye, the lucidity and transparency of desert light confounds the sense of distance, by modifying the visual cues from one's normal expectations. It takes experience to "read" the desert, because sharper edges between form and background persist for greater distances, owing to the clarity of the light, causing them to be consistently underestimated.

6-15

6-17

6-16

The variety of greens seen when a mountainside is scanned indicates the state of growth of trees and shrubs. The bright green of a rhododendron's new shoots against its deep, lush background, the yellowish young greens of emerging tendrils, are associated with new vegetation.

A reading of shells shows that for each species, individuals follow a rough pattern, which no single shell follows precisely. While a shell can be classified and described by other morphological features, size, and shape, its very individuality is assured by coloration. It is evident that its color patterning and structure are simultaneous processes. Within calcifications distributed radially as a pattern of growth, colored markings will occur as frequencies, changing scale in conformity with the increasing size of the shell. Patterns begun at the center grow in size and area to the edge, simultaneously layering calcium and coloration. To the eye, these markings can look decorative or may more fundamentally reveal the structural process itself. By conforming to surface variant, color in this case reveals form. Patterns *assign* forms in nature; they signify.

6-18

On the color-form issue nature is clear, but she does appear to have it both ways. Color both conforms and transforms. Or perhaps it transforms by conforming. While the patterns of a zebra seem at first to be disrupting its volumetric surface, a closer reading will show that the blacks and whites also encase and delineate. Defining a haunch or a neck, marking a change in direction in the juncture between leg and trunk, delicately changing scale to conform to the narrowness of a leg, the zebra's stripes are a result of its architecture. Its surface is a topological diagram, an expression of organic structure. This subtle interaction is greater than either alternative.

Nature shows us intrinsic order. What we tend to observe in its surface manifestation is decoration, but this is a semantic difficulty. Visual language is clearer; natural color seems to be an organic expression. Nature makes surface display that is not superficial.

Various and marvelous as these effects can be, they seem even more so when considered as part of parallel systemic processes. Surface is not additive, but intrinsic; color is integral with form. While these manifestations of invention and consistency must be one of the most significant aspects of nature, it is also the least studied.

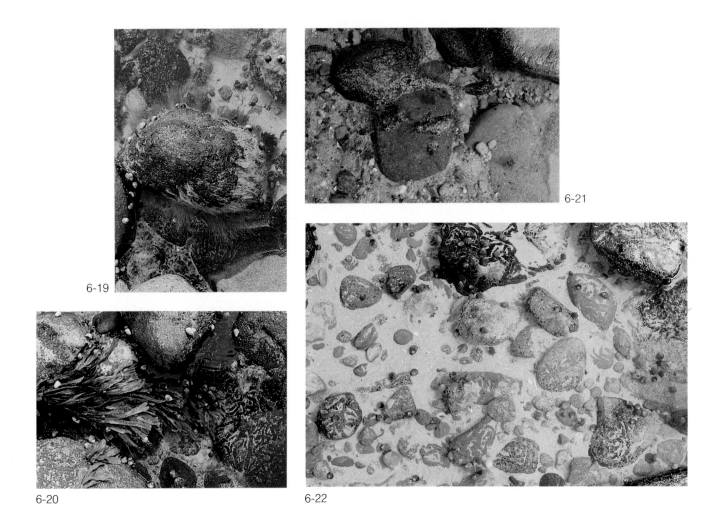

6-19

6-20

6-21

6-22

A tidal pool at Martha's Vineyard shows an abundance of life on an environmental microscale, expressed by its colors. While stone, seaweed, algae, insects, and sand each manifest a distinctive coloration, the whole appears integrated in effect. If the variety is visible in the hierarchies of scales of textures, microstructures that modify each surface appearance, integration may lie in the complex effects of light. Sunlight intensifies all surfaces consistently, while the water patterns absorb and differentiate, casting a veil.

Natural color is complexly structural, in its close association with light, its combined effects with hue can be richer, and at the same time subtler than human artifacts. To those who use their eyes, encountering her directly, nature is a source better than any book that can be written, on how these variations are manifest.

7   Chiaro e Scuro Inverted

Leonardo da Vinci characterized chiaro e scuro as an invention in the art of painting, to depict the illusion of volumes in space.[57] His conception referred to the gradation effected from the light to the shadowed side of a three-dimensional object, as a development of a range of grays, arising from his observation of their placement or distribution relative to their surfaces. That both the term and the visual structure it defines have endured is testament to the "truth" of the discovery.

But the relationship between the perceiver and the perceived is not as simple or direct as it might have seemed to be. What the eye receives as the intensity of light, and its absence as shadow on the retina, does not directly correspond to the illusions obtained by painting and drawing. Rather, the visual system makes a translation of the enormous variation in the degrees of differences between the lights and darks which exist in the environment, to patterns that can be perceived. The human eye is capable of discriminating between about ten million differences in levels of luminosity, yet with a very limited light/dark scale, the entire visual world can be reconstructed, or simulated.

When Viking One landed on Mars, its image was transmitted to Earth electronically. With a numeric scale reduced to a limited number of degrees of intensity, the sensory reception, pixel by pixel, in vertical lines on the monitor, was constructed to image; and a pattern of lights and darks revealed the forms of rocks, planes, sky, and Martian surface, and an entirely new landscape was made visible! This scientific event as poetry was the result of the capacity of visual systems to reconstruct complex reality, whith a minimum of means. Had he been witness to it, Leonardo himself would probably have been moved and impressed by the application of his formula.

The context of the Martian environment, as well as its individuated forms, were translated and transmitted by a scale of luminosities, represented by incremental levels between light and dark. In fact, these steps could not represent actual, measured reflectances, but had to represent averages between surface luminosities in the entire visual field. When numbers were translated into light intervals, the reconstituted image simulated the total effect, which was it-

7-1

7-2

self the resolution of ratios between individual areas or surfaces. Photographic film also works with ratios, its variable light sensitivities are keyed or graded, so that the same image can result from different time frames in exposure.

The significance of technology to an understanding of the visual system, is that the process in those cases, *simulates by reduction*. I think that the visual process may be similar. In order to maintain coherence from one environmental light context to the next, the visual system processes the stimuli, however great their differences in range may be, to a pattern similar enough to be consistent to the eye and meaningful to the brain.

For example, aware that the eye requires a minimal stimulus to reconstruct "reality," an artist can suggest that a line read as form. Even if incomplete, a boundary describing a volume in a field will be filled in by the eye, to create the figure. In a tonal drawing very few grays, if they are organized coherently, can result in the illusion of solidity.

A comparison of individual styles and media can demonstrate this further.

Leonardo developed his chiaro e scuro as a rich and full scale in his sepia and chalk drawings. Lights and darks, as well as lush intermediaries, describe the range which his analytical eye observed in nature. Comparing a Leonardo drawing to one of Ingres, the sharper, more linear characteristics of pencil or silverpoint reflect Ingres' vision. A minimal artist compared with Leonardo, Ingres implies more volume than he describes. The distinctive clarity of a sharp line, or a rich range of values express diverse reactions in each artist to objective reality. But by differing modes, each achieves verisimilitude by means of reduction.

Pattern making in visual perception then, begins with light and shadow. In previous experiments we saw how relative the appearance of an angle is to the distinctions between lights and darks, or the value components of color. Light and dark is the fundamental building block of form, because the contrasts between them can determine boundaries, or subtle gradations visually organize as surfaces.

7-3

7-4

The Pyramid Problem:
A Visual Ratio

To understand the differences perceived in light and shadow between four sides of a square pyramid, the sequence in chiaro e scuro was compared to a scale of gray papers. If a specific pattern of light and dark reconstitutes form on a two-dimensional surface as a drawing or painting, then what proportion between them accounts for the visual experience of the surface per se? Is there a relationship between steps of gray in the four perceived, and its height? Chiaro e scuro is a *visual relationship,* a balance achieved by the eye and brain in response to the light intensities that determine contrast in the environment.

In a project each student constructed two square pyramids, identical in their size and proportion, 9 inches square at the base, and $5^1/_4$ inches from base to apex. One model was left uncolored, the white of the mat board, while the other was colored by covering each of the four triangular faces with papers cut from Color-aid grays. A scale of eighteen grays between a white and a black accounted for all possible variations in light and shadow.

Using the natural light of the studio, with its variety of angles of incidence and intensities of light, rather than setting a single experimental condition, each person pinned the two models side by side against the wall. Reading a single pyramid with its apex at eye level, from a distance of about 10 feet, the observer experienced alternation in its appearance, either as a volume or as a square intersected by two diagonals. With a slight change in position, at eye level and with the observer's head oriented midway between the two models, both pyramids attained equal perspectival distortion.

At first each model was studied from a direct, frontal position and the precise distribution of light and shadow on its four triangular faces was analyzed. As each was located at a different angular relationship to the light source, a window along the axis of the studio, or a skylight above, the pattern differed. The steps in value or brightness gradient observed were noted as intervals of white and gray papers, selected from the scale of eighteen. When four steps were found which simulated the boundary conditions produced by light and shadow on the model, they were cut from the papers at facsimile size, pasted on a two-dimensional surface, and pinned to the wall adjacent to the pyramid. When compared, the flat paper analysis and the three-dimensional appeared similar in pattern.

The incidence of light set the pattern; the lightest plane faced the source, with the darkest triangle of the pyramid opposed. The pair of remaining triangles appeared equal to each other or not, depending on reflectance from surrounding surfaces. Having simulated the contrast effect, we were ready to test illusion with reality, as we had before, with the study of the 90 degree angle.

Four additional triangles were cut from the papers chosen for the simulation, but placed now against the second model. Assigned in reverse order (that is, the darkest gray against the lightest face, etc.), the four values should not compensate precisely for the chiaro e scuro pattern, according to the geometry of brightness discovered in the angular experiment. Indeed, they did not; the pyramid appeared gray, and its diagonal angles softened. It was shallow compared to the white model adjacent, its altitude appeared diminished.

The next objective was to find a match between the four triangles where they appear equal in value. By trial and error, papers were tested against the gray model, until the contrast boundaries between the triangles were diminished to imperceptibility. These four steps in value, compared with those originally chosen, were greater almost by half.

When the four triangles appeared as a single gray, the pyramid lost altitude entirely. Observed frontally and binocularly, it appeared to be a square plane, fitted to the wall. By comparison with the white model, it was flattened, even when the two were observed from different positions in the room, and from angles of 45 degrees. A discrepancy was observed in the tilt of the diagonals between the triangular planes. On the gray model these were distinctly straighter than those of the white pyramid! The effect of angular relocation, a function of diminished contrast, was sustained from different vantage points. It was a surprising confirmation of the relativity of form.

After the faces were covered to flatten the pyramid, observations were made of its appearance in rotation. 7-5 shows a gray pyramid, cancelling the light. 7-6 shows the object turned 90 degrees clockwise. The effect perceived in this position is that of a diamond shape bisected diagonally, with four grays appearing as two pairs of equal grays. Continued 90 degrees clockwise, the square is inverted from its original position, and the four grays are illuminated so that the lightest gray reflects the most light, the darkest gray, the least. 7-7. With the steps in value increased geometrically, the volume appears overly accentuated. To some observers the effect disintegrates the volume, forming once again four separate triangles.

We characterized the changes as follows:

 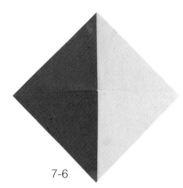 

7-5   7-6   7-7

Stimulus	*Number of Triangles Seen*	*Spatial Effect*
Four values	None	Pyramid flattened
Four values	Two+two	Pyramid bisected
Four values	One, two, three, four	Exaggerated altitude or four triangles

Interestingly these changes were sustained even when the pyramid was removed from the wall and hand held. By reestablishing the original angle of incidence in a different spatial plane, the pyramid was turned 360 degrees and the effect was observed. When the four grays appeared equal, the object seemed to contract in one's hand. In continuous rotation, as the four approached maximum contrast at their boundaries, the object seemed to expand. This magical illusion persisted, despite the fact that the object's dimensions were experienced tactilely.

We tested the effect next under a different level of illumination, to see whether it would persist or change. Taking the gray model outdoors in bright sunlight, where light intensity is a thousand times greater than that in the studio, we reestablished the position of the object with respect to the angle of incidence of light. To our great surprise, despite the vast discrepancy in context

of the environment, the illusion that had been obtained in the studio "worked" outdoors. We experienced the flattening, bisection, and expansion of the model as it was rotated in a plane.

Significant to the appearance of volume is the angle of incidence, rather than the intensity of light. The proportion in contrast persisted on the pyramid, despite great differences of illumination in the environment.

Applying this discovery to my own work, painting on three-dimensional reliefs, I found that once a pattern of values has been established with respect to a particular angle, the appearance of the work retains consistency whether it is viewed in the natural light of the studio, or in a gallery, under artificial light.

Modules such as the pyramid applied as surfaces to architecture, can be colored to respond to ambient light. As a perceptual building block, the constancy in light-dark proportion is a significant aspect of form, providing visual continuity. While its appearance might fluctuate with shifts in sunlight, a building's surface can be designed to change according to its orientation or location in the environment.

If the pyramid ratio persists despite appreciable differences in environmental brightness, then its constancy must be a function of the eye and brain. The visual process may compensate for the differences in the grays, so that they remain in proportion to one another when there is an increase or decrease in the amount of light. Thus *relationship* is significant, or the constancy of form is a matter of *proportion* between light and shadow.

With respect to the changes in perspectival location of lines, it seems in this case that a line is a threshold, conditioned by the contrast, or its lack, between areas of intensities. If edges are a function of the distinction between light gradients, then their softness or clarity also define spatial difference or distinctiveness. But what accounts for angular relocation, or loss of perspectival distortion by loss of contrast?

Having established a formal system with value, or the gray scale, we proceeded to test hues on the pyramid. In one experiment we used the gray pyramid itself as a field. Choosing a series of hues from the range of 202 colors, we found four which, modified by their value component when each was placed adjacent to the next, appeared to have the same step relationship as the four

grays to which they were compared. Four small paper triangles of different hues were placed against each of their gray equivalents on the four triangular faces of the pyramid. These smaller triangles were fitted adjacent to its apex and observed frontally, so that a visual pattern of four colors appeared as a small square field fitted in the center of the pyramid, with a ring of four equivalent grays surrounding them. When this match was compared, the resulting illusion was that of a four color film overlapping the form's apex, a three-dimensional transparency. Seen in rotation, if all four grays appear to be equal, the effect of a single film results, consisting of four hues. When all four grays appear maximally unequal, the four hues related to them appear spatially segregated. Their relative locations seem to be determined by color-spatial activity, so that if a red is used in the cluster for example, when illuminated by rotation it will move forward, disrupting the grouping. Value serves to contain or equalize the spatiality of hues, but fully saturated hues exert an influence on formal patterns. There is a tension between the logarithmic system of brightness and the dimensions of hue.

With the value ratio established, a variety of patterns with hue were tested against the pyramid. Bright and dark versions of a blue applied as sectors against the triangular faces for example, effectively bisected the pyramid. Games were played with other hues as spatial transpositions over its surface, sectioning and redefining the pyramid as a form.

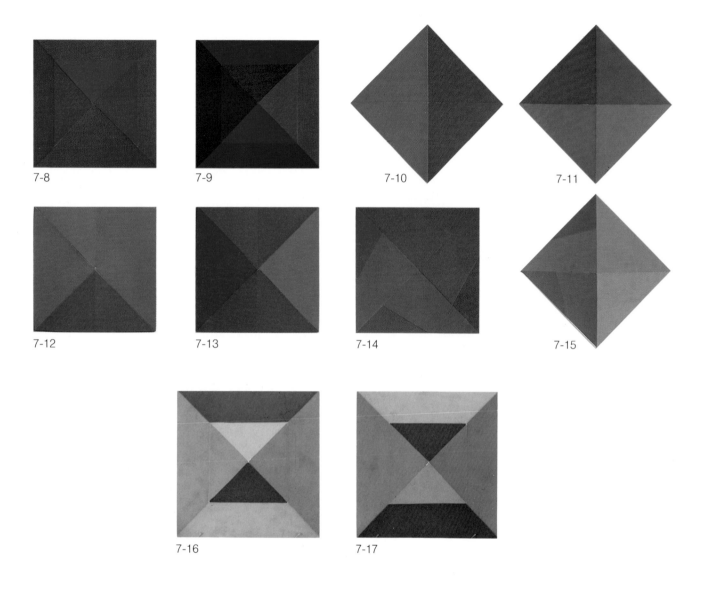

7-8

7-9

7-10

7-11

7-12

7-13

7-14

7-15

7-16

7-17

Following these rigorous, analytical studies, I encouraged free problems. How many appearances can a pyramid have? The most compelling solutions were those that referred to the geometries of form and light, by playing with colors patterned according to these underlying visual orders. Having experienced the analytical phase of the problem, individuals seemed to prefer not to be arbitrary, but to relate their inventions to the pyramid's structure. Circles or curves integrated into triangular fields were more often geometrical than "free," and straight lines predominated. Of the hundreds of individual solutions, there have been no repetitions. But it does seem that the visual thinker has been influenced by his or her experience to accept constraints even with freedom. The aesthetic decision has been to transform by conforming.

8 Color and Visual Organization

Hues cluster. This effect may be a function of the propensity of the eye and brain to make primary distinctions between colors. When grouped in an expanded visual field, individual hues tend to organize so that reds are visually connected to reds, blues to blues. It may be that the combining of hues according to like kind is the result of an opponent-response system in vision.

When hues are separated spatially they group by proximity, warm to warm, cools together; and oppose one another as clusters, warms against cools. This propensity to constellate occurs despite their actual location in a field.

In a group of experiments the relationship of color to Gestalt principles[58] of form and organization were explored:

Principles of Organization and Articulation
Configuration
Unit formation and segregation
Proximity
Similarity and equality

7-18, 7-19, 7-20, 7-21, 7-22

7-23, 7-24, 7-25, 7-26, 7-27

7-28, 7-29, 7-30, 7-31, 7-32

7-33, 7-34, 7-35, 7-36, 7-37

7-38, 7-39, 7-40, 7-41, 7-42

Colors can describe properties of configuration, self-containment, and unit formation in their propensity to group and segregate. What occurs however, when color is only an element among others in forming units? In a three-dimensional field is the size or proximity of objects a more important factor than their color to an observer? To study these issues I colored cubes of different sizes and placed them in a black field, in order to understand their interaction and to assess the relative importance of these factors to the eye in organizing them as objects in space.

In a space-frame[59] 4 feet by 4 feet by 8, a black box open at three sides, five cubes of different sizes, covered with red, yellow, and blue papers, were placed. They were positioned to appear equivalent in size when visualized through the box's open longitudinal end.

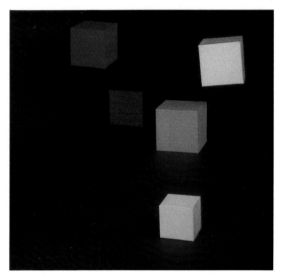

8-1

In 8-1 the small yellow cube near the observer, and a blue placed higher in the same lateral plane relative to the eye, are equal in size. The largest cube, colored blue, has been placed at the far corner of the space-frame. While it is 8 feet distant from its nearer and smaller equivalent in hue, it appears somewhat smaller to the eye. With monocular observation the size/distance relationship is made ambiguous, and similarity of color is more significant than proximity in estimating the spatial placement of the two blue cubes. A larger red cube is placed in the frame at a distance to the observer where it appears equivalent in size to the small yellow cube. Finally, a second yellow cube, larger than the first, but smaller than the red, is placed within the long axis of the frame to appear equal in size with the others, when observed monocularly. Without depth perception all five cubes now appear equivalent in size, as colored objects floating in a black field. Since clues to their rela-

tive size or placement are reduced, the most coherent reading is obtained when they are grouped according to their color.

The yellows appear to connect with one another while at the same time, they tend to anchor to the centrally located red. The two blues identified with one another, have a more passive appearance, receding from the cluster of warm colors. Thus the two smallest cubes, one yellow, the other blue, which are closest to the observer, appear to be maximally separated in the space-frame. When the blues are linked visually they tend to recede, while the two yellows appear to advance. While monocular observations varied, some individuals reported a spatial clustering of two distant blues and two near yellows as the branches of a "Y" around the red. Most readings of the cluster were made according to the proximity of colors, rather than the true location of the cubes in space. Stereopsis resolved the

8-2

illusion. Size, placement, and depth more rationally disposed the elements, while disparities between the colors increased ambiguity in their grouping. Color is the dominant factor of organization only to the *monocular* observer.

In 8-2 the cubes are placed to be observed from the lateral axis of the rectangular field. A large red cube is placed in the frame at its far corner, relative to the observer. Two small cubes, one yellow the other blue, are placed in the same lateral plane, close to the edge of the frame and to the observer. The yellow and blue cubes appear distinctly smaller than the red, but not equal to one another. In this case the frame of reference of the field is an influence on the reading of the relative sizes or distance of the cubes. The yellow, placed below eye level, can be "read" in relation to the edge of the frame and appears closer or larger than the floating blue cube above. A fourth cube, colored yellow and intermediate in size between the blue and the red, is placed

against the rear wall of the space-frame. The discrepancy in placement between the two yellow cubes of different sizes, causes the larger to appear to float. Their proximity in hue however, connects them, and the constellating effect around the red can be observed again. When size and placement are restored as clues by binocular observation, the constellating effect of color is secondary to the formal factors in the spatial organization of the cubes in the frame.

In the two instances described there is a strong tendency for the yellow and blue primaries to constellate around the red. While in the first experiment the red appeared equivalent in size to the yellow and blue cubes, its central position in the cluster supported the effect. In the second it was enhanced by size. But in both cases the true location of the red was belied by the insistence of its hue. In the triad of primaries the red exerted the strongest spatial effect, making it the pivot for the cluster.

8-3   8-4

Another example of constellation was obtained with four planes of equal size, one yellow, one blue, one red, and one purple. These were placed in different planes in the longitudinal space-frame, and observed from its open end. Again, it was observed that the relative displacement in the depth of the field caused ambiguity in their reading. To the monocular observer they can appear as four colored planes of unequal size in the same spatial plane or four equivalent planes in different locations in depth. Their propensity to cluster occured when they were observed monocularly, and off the central axis of the field. Displaced in depth they appeared to overlap.

In the illustration, the blue and yellow planes seem to share a common boundary, while the red and purple are seen adjacent to one another. The yellow, located at the greatest distance in depth, is smaller and brighter than the other hues. Attached to the edge of the blue plane or overlapped by it, the yellow seems to be located in front of the blue. Below the pair, the red and purple planes assume a horizontal position relative to the monocular eye. The red

appears appreciably closer than the purple, despite the fact that it is situated behind it in the frame.

When all four hues are placed in depth so that the red and yellow appear to share a mutual boundary, the red assumes the pivotal position in the constellation of the group of four that had been previously observed with three. While in these experiments the purple plane was located closest to the eye, it did not appear to be. Despite its greater size and its location in the field, purple appeared as the most distant color due to its relative desaturation in the context of brightly saturated primaries.

Next two spatial dividers are placed in the frame. A horizontal line subdivides the black square of the back wall of the box. A vertical line intersects this and continues as a divider along the longitudinal axis of its floor. The black space now is subdivided visually into four rectangular sectors, which delineate a new frame of reference for the four colored planes. When the yellow, red, blue, and purple are moved back and forth within the longitudinal field, each can be seen in relation to its linear frame of reference. The ambiguity perceived before in an undifferentiated black field is eliminated, and the size/distance clues are restored, even with monocular observation.

This series of experiments show that color can have a significant influence on the perception of objects in space. The frame of reference is the factor least influenced; that is it significantly determines location. But size, distance, proximity, overlap, contiguity are influenced by hue, and when observed monocularly, color can be the predominant factor in organizing groups of objects in space. While stereopsis restored the importance of the formal factors, color remained as the signal or focal or localizing attribute of form.

Two-dimensional studies are less ambiguous. In a study using basic shapes, squares, circles, and the psychological primaries, red, yellow, and blue, psychologists found that observers made distinctions in the grouping of colors, rather than the shapes, when organizing the pattern.

The studies that follow, conceived of in three dimensions, play upon the interaction of color and form in space or within a frame or reference.

The Multiple:
A Creative Game

First-year architecture students experimenting with color in three dimensions made a multiple with basic elements. The objective was to devise a game of wooden blocks, the same size, cubical in shape, with one surface truncated at an angle.

Blocks cut at five angles, repeated five times with five colors, a total of twenty-five elements constituted the system. They were placed in a square base. These limitations permitted a study of color, angle and light and shadow as integrated pattern, by moving them on the plane. Each student selected his own group of five colors, with a range of preferences expressed for analogous, harmonic groupings, for complementary pairs, as well as arbitrarily chosen clusters of hues. The top surface varied in angle, from 90 degrees through three intermediate steps to a 45 degree angle, so that a profile emerged when they were aligned in sets of five. Colors were assigned, so that blocks of the same angular profile were the same hue, or alternatively, the same color appeared once on each of the five angles. When combined the groups appeared as a multicolored and profiled system, affected by light and shadow.

8-5

8-6

8-7

8-8

The objective of the game was to make a "good" group, without preconceptions as to which factor, color, or form would predominate in effecting decisions. It was more interesting to leave the questions open and to observe the individual's resolution.

When a color was assigned consistently to an angle, integration of profiles with color was attained in combination. If each block having the same color were a different angle, then the two variables worked independently as systems. The choice between the two options resulted in a different strategy in game playing.

By reducing the two variables to one, when an angular block had the same color, groups of five could be repeated in rows of five to fill the square field. This opened the game to the possibilities of combination. The same linear sequences could form a spiral or meander. More rigorous and complex patterns were tried. For example, a fourfold symmetry occurred when clusters of five identical colors were placed as cross configurations within the field. 8-7.

Two fourfold groups were contained in the square field when formed by groups of five similar colors, or by groups of three placed horizontally, with two contrasting colored blocks as the vertical element, illustrated. When the field was divided diagonally by the same color, the five blocks determined a mirror for the remaining four groups of five colors. 8-5. These were distributed in one case so that analogous colors in mixture appeared as a gradation over the square area.

Separation of the color system from the angular one, inferred a sequence in color or in profile, but not both. There was a tendency to group assemblages designed this way more intuitively, or as in the case illustrated, where the color is logically arranged, leaving the profile to chance. The play of light and shadow in this example can cause the appearance of more than five colors in their variation. 8-10.

By not making assumptions about how the game should be played, whether according to a rational strategy, or intui-

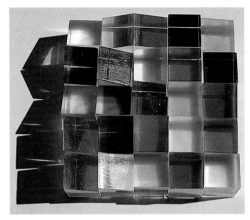
8-9

8-10

8-11

tively, we confronted many issues during the process of assemblage. How many groups are possible with these twenty-five elements? Is angular repetition or color more significant to the eye? Can there be random groups with this number of elements? Is "perfect" order possible? Or boring? What are the criteria for "good" groups?

Since the blocks could be rotated to face a new direction, once placed, the same angle could repeat as a mirror symmetry, producing peaks and valleys over the field or a multiplicity of skylines. A solidly organized spiral with a large cluster of color at its center was enlivened by the disparity in its angles. The design of the system had offered only two options; therefore an implicit preference for order predominated in the solutions. On the whole, perfect sequences in angle and color gradation were considered too simple or dull, like playing repeated scales on the piano with five notes, a minimalist's preference. Those considered to be best by most students were solutions containing aspects of both order and randomness.

While architecture students tend to be form conscious and their thought processes logical, color sensitivity increased during play to determine pattern. Focus on color was combined with chance combination of angle and more arbitrary groups were tried. Overall the experience showed that the basic game of building blocks could be more varied and sophisticated than had initially been imagined.

In a Plexiglas version of the model, which I did in collaboration with Velizar Mihich, squares of opaque colors are placed beneath cubes of clear plastic. Light appears as color on the precisely cut and polished angular faces, a function of inner reflection of the clear material. In this case colored chips and clear blocks can move independently, and the possibilities of combinations are increased. 8-9.

8-13

8-12

8-14

The Multiple:
Devising a System

The multiple can be a means of spatially distributing colors. The common wooden spool was chosen as a base for thinking about variations of three colors on a number of identical units. Its symmetry and cylindrical form offer a neutral field for color placement; it has a multidimensional aspect, top, side, bottom, with features of reversibility.

I assigned a problem using three colors distributed on thirty-six units, so that the possibilities of clusters would add variation, three groups of twelve, six of six, and so on. To visualize their placement on the three-dimensional surface, one can begin by lining up the elements as spatial modules with a constraint that no cylinder may identically repeat the same pattern of three.

In the example illustrated this aspect of the problem was solved logically. If the spools are in a linear sequence, the three colors can be painted as a series of intersecting planes. The effect is obtained by the selection of a light green, a dark blue, and a grayish green, distributed as patterns through three lines of twelve units. Two colors appear in each case to slice through the spools, beginning on their circular surfaces, and continuing through the vertical cylinder, to the circle on its opposite side, reiterating the form's symmetry. Gradually, these colors cut through the sequences of twelve units, appearing to move through one another. The gray functions in this example as a perceptual middle mixture between the two hues,

8-15

8-17

8-16

so that in the arrangement, the transparency effect is used to advantage.

Once the system was designed and executed, games were played. Patterns can be observed in combinations; a three-fold symmetry arises when units are clustered in a close-packing configuration. When integrated by a square base possibilities are determined by turning cylinders relative to one another. Then their linear patterns form diagonals, and color interactions depend upon the reading of hard, soft boundaries or "transparencies."

In this solution, the balance between the numeric possibilities of the combinations of thirty-six units, and the perceptual logic of their color distributions, have been particularly well integrated.

The Multiple as Module: Toward the Design of Buildings' Surfaces

Multiple units can be used systematically as a factor in pattern formation, or they can be freely applied as modules, in thinking about the design of architectural surfaces.

A change in material offered this possibility. Mailing tubes of fiberboard are available in a variety of diameters. They can be cut to any length on a bandsaw, at any angle. Their inner walls, as well as the outer surfaces can be painted, and the thickness of the rims become a third variable. For the distribution of three colors therefore, it seemed logical to clarify and articulate these three formal attributes. By cutting the cylinders at acute angles the interior of the tube was opened to the light, disposing the chiaro e scuro patterns of the repeat form to variation. To do less with color was to accomplish more.

8-17 shows thirty-six units, containing six angles, a six by six distribution of three colors. The analogous relationship in hue confers an integrated effect, and the group has visual coherence from any vantage point.

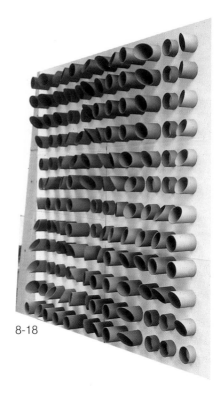

8-18

As it is, the panel has a multifaceted effect. It changes appearance and with a shift in the observer's orientation, new relationships arise. When the inner/outer colors interplay they make the chiaro e scurro pattern more complex. As the angle of the ambient light shifts the patterns are modified, the hard-edged profiles are distinguished against the light background by their dark contrasts.

There is a rich interplay of the dimensional surface; a function of angle, light and shadow, color, the time of day, the position of the assemblage, and the vantage point of the observer. A statement or commentary on these aspects of change is directly expressed here. If these units had been repeated extensively over a building's surface, the rhythmic possibilities of the curvilinear profile would have had to be considered. A facade responsive to environmental changes to fluctuations of light and atmosphere is intriguing to contemplate. Such experimental models as these are ways of visually integrating color with the extended architectural surface.

8-19

The idea has been applied in 8-18 to a wall panel. Composed of 144 units, the elements are disposed in a grid, twelve by twelve. The three colors signify the interior/exterior/edge dimension of the tube, in all possible variations. When assembled, the color pattern is fairly randomly distributed, but the differences in height and angle of the individual modules read as a curved surface when they are combined. While this effect had not been anticipated, it arose as a result of placement and repetition of the angular gradient. The panel at this stage of development could itself become a module. Repeated, the curvilinear effect is potentially a wave pattern, which would be more in evidence with a greater number of elements in the system.

8-20

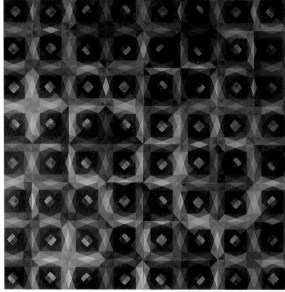

8-22

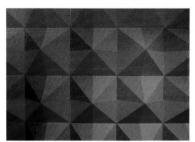

8-21

Modular Patterns: Walls

Installation 1: Harvard

A colored pyramid is the module repeated as a grid in a large-scale study for an interior wall. Designed for the second floor landing of the stairwell of the Carpenter Center, Corbusier's building at Harvard, the temporary installation remained in situ for a period during which fluctuations in indirect daylight and artificial illumination were observed and recorded. Intensity changes on the pattern were subtle in daylight; artificial light influenced appearance due to its incidence from above. Note the rotation of the single module and its effect on the total pattern.

Installation 2: UCLA

Here a single pyramid, designed to change appearance by rotation in light, was photographed in sixteen different positions. The influence of light on its surface, once recorded, was repeated by photographic print. The sixteen units repeated four times constitute the total pattern. Complex effects of light and shadow interact with the original unit to form large-scale overlapping patterns.

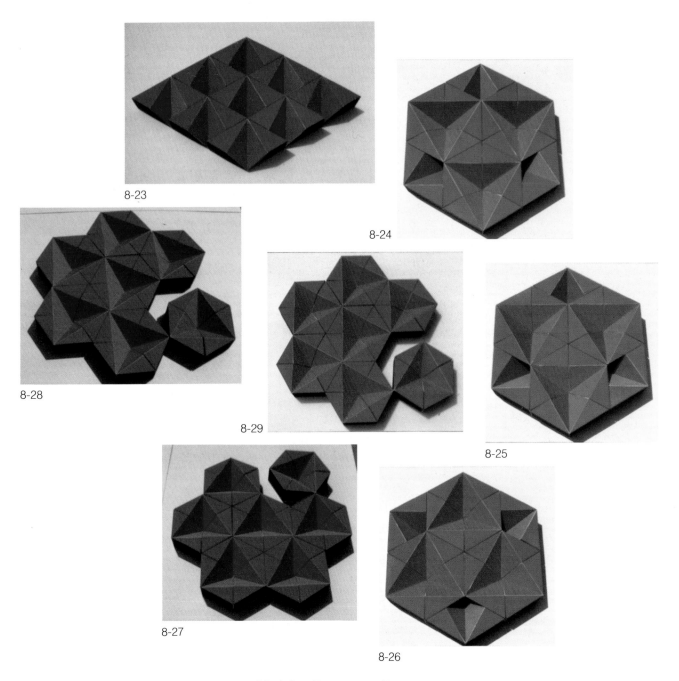

8-23

8-24

8-28

8-29

8-25

8-27

8-26

Modular Pattern as Repeat

A four-faced unit is formed by folding stiff paper to make a three-dimensional module. A gray, a light yellow ochre, medium light red, and blue-violet were selected to cover its faces. Each module contains a combination of two hues, with a total of six possibilities in combination, as seen. On a plane the module can form a surface and the units can be combined as three different

8-30

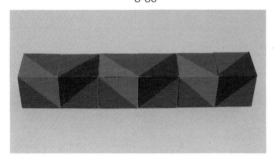

8-31

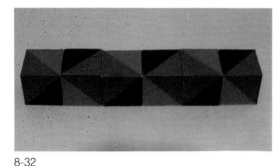

8-32

8-33

The Module as Unit in Film Animation

The module of a cube offers the possibility of covering six faces four ways. A distribution of the colors blue and green and their variations in value evokes comparisons with light and shadow. When combined as units the faces of the cubes cluster differently as patterns, some forming diamonds, others chevrons. Illustrated here in small scale, a group of 2-inch cubes show some of the surface variations. Similar cubes 3 feet in size were assembled in a variety of possible combinations and recorded by animation camera. These sequential changes, frame by frame, produced motion effects in a film that shows the variations possible in the real time of one minute.

hexagons, based upon the color variants ochre/pink, ochre/violet, pink/violet, pink/ochre, or violet/ochre, violet/pink. In reverse order the groups change their color by the influence of light/ shadow. They can also form three tesselated groups of seven small hexagons, or three parallelograms as the logical consequence of their structure.

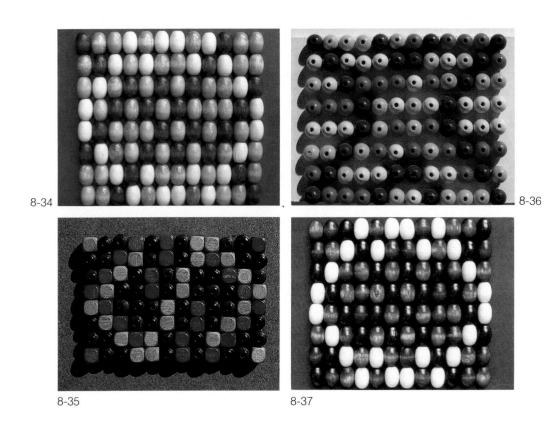

8-34

8-36

8-35

8-37

Pattern-Making:
From Sequence to Consequence

Using colored elements of the same size and shape, patterns were constructed based upon the logical combinations of four. Wooden or plastic beads can be strung to make a linear sequence. For the possibilities of grouping four, a potential of ninety-six are placed in a line. The options are to proceed numerically with four colors, as sequences of 1:2:3:4, then 2:3:4:1, etc., then repeating the occurrence of single colors, until several possibilities are tried in a single line. An alternative strategy is to proceed intuitively, using the four colors as they combine aesthetically.

The long line of strung beads is then adapted to a square or rectilinear field, where it forms a unit pattern. While no repetition has occurred in the stringing process, except for mirror repeats of four colors, when the line is fitted to fill the field, unpredictable clusters of the four colors will result. Some do repeat, by inversion. Chance alignment in the horizontal caused by the placement of the string as a spiral over the entire surface, or by its vertical repetition as a meander, adds a novel variable.

As a consequence, the rigorous alignment of groups of four colors can result in unpredictable or asymmetrical patterns in the field. The visual statement richly asserts the possibility of order and variation with four colors, whether the unit was initiated logically, or intuitively, by "preference."

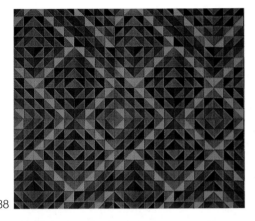

8-38

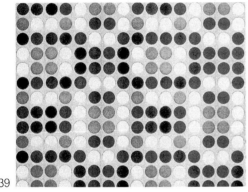

8-39

Unit to Module to Field

In another instance, the sequence of four colors was done with colored stickers. As a first step toward the organization of a field, four colored circles were aligned in vertical groups of four, and placed adjacent to one another in a sequence. When all possible groups of four were placed, the entire group formed a long horizontal, which we called a "color sentence." The best clusters of four were then selected as phrases from the sentence, to work together aesthetically. The repetition of the four colors formed a grid with the white background. When the colors were close-packed as a parallelogram, the whites worked as small interstices. If the colors were repeated equidistantly against the white, to form a vertical/horizontal grid, the white functioned visually as a fifth color. Once a coherent or visually pleasing unit had been derived as a phrase, the remainder of the "sentence" was discarded. The group of twelve, sixteen, or twenty colors selected by this process, was then repeated by color xerox. This unit was

used in a variety of ways to fill a rectangular field. By rotation, mirror repeat, or glide, the single cluster gave rise to a pattern that could be expanded spatially.

Pattern has a role to play in extended architectural surfaces. For the continuous, unaccented, long corridor, which is the experience of passage from entry to gate in airport terminals, patterns that rhythmically punctuate would offer relief from the prevailing boredom and sterility of the environment. A surface pattern can mark or measure progress by a change of size, scale, or color; it need not repeat endlessly. The demarcation of distinct zones by color can alter monotonous pacing and serve to locate the traveler in his passage.

The issue of scale becomes a design consideration, since the shifting of size or placement in a pattern can significantly alter the perception of space and distance. This aspect of the extended architectural surface bears study and can

8-40

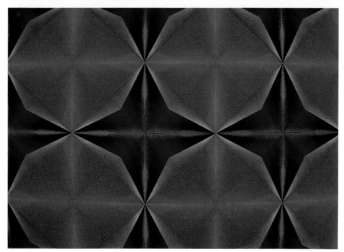
8-41

entail the graphic means of the computer. In addition to size/distance changes, the effects of scale in colored patterns can also cause additive mixtures. When patterns are very small, independent colors begin to mix or to assimilate perceptually. As the computer is electronic, its color source is light. Additive color was studied in its effect as colored lines as gradients on a Chromatics computer, which offers a visual resolution of several million pixels, or dots of color. When a pattern is generated numerically, its translation to a visual module on the scope can be repeated with the touch of a key. Once established a pattern can be enlarged or diminished in a matter of seconds. More significant to the issue at hand, the repeats can be changed in size and area and visualized, so that a building facade designed with patterns as a surface feature can be studied with respect to its effects upon scale. Simulations of the perception of surfaces at relative distances from the "eye" are made possible. The computer offers an encapsulated time frame to inculcate and develop simultaneously the sensory and mathematical sensibilities which are requisite.

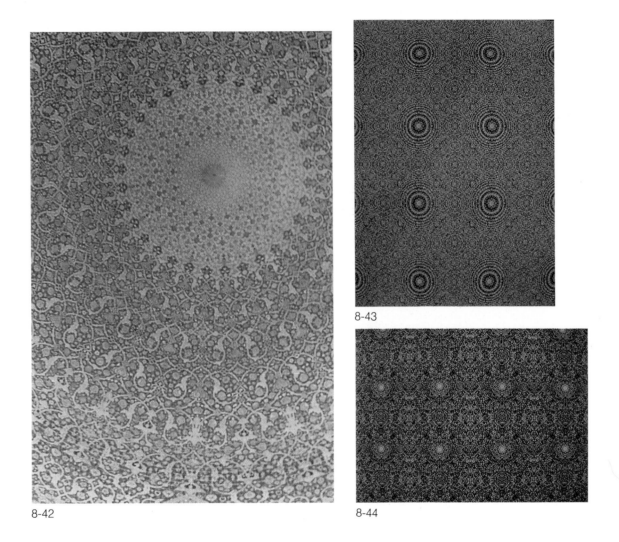

8-43

8-42

8-44

The Persians evolved their systems over generations. The processes of structuring, evaluating, visualization, and variation are the evidence of a traditional art form that develops over time. Arguments can be made for the gradual evolution of such forms, since judgments are not made instantaneously but take into consideration such issues as context and time. Environmental factors such as light, the materiality of the color, the methods of facture, deviations from exact repetition, are significant influences on how Persian surfaces are experienced.

However, as the issue is relatively new for contemporary architecture, this interim step, the testing of scale effects of surfaces by computer simulation, can be valuable. In the process, we encourage the development of that human hybrid, the gifted designer, a visualizer who is equally at home with numbers. So were the Persian architects and craftsmen.

8-45

Patterns as Projections in Space

The constellating effects of color, by
their localizing, spacing, and dispersion
can be visualized as surface patterns.
Can this process be extended to the
depth of the three-dimensional field?

The distribution of colors in depth is a
kind of three-dimensional configuring.
While Mondrian rigorously maintained
the integrity of the picture plane, his
two-dimensional paintings were the
model for a three-dimensional analogue.

Can a basic exercise predict a process
encountered later, the visualization of
spatial entities in the complexes of urban form? Is it possible to conceptua-
lize with color directly, as a sculptor is
accustomed to when modeling in clay
or wax or carving marble? Is there a
link between the visual and tactile
sense, so that the plasticity obtained by
Mondrian in two spatial coordinates can
be realized in three?

A cubic field was defined as a unit by
the distribution of transparent plastic
planes in space. With the use of a
square wooden plane for a base,
grooves were sectioned to the size of
the plastic's thickness and repeated re-
gularly as parallel cuts. Five transparent
square planes of Plexiglas were slid into
the grooves, at 90 degree angles to the

8-46

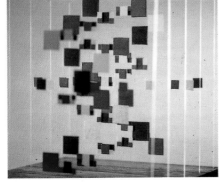

8-47

base, providing parallel surfaces. The cube had two basic orientations to the observer, front and back, with secondary views from two sides and above.

Students were challenged to think of the unit as a total field. The psychological primaries, squares of red, yellow, and blue were to be dispersed throughout the volume. All three hues could appear on each of the five transparent surfaces, or each could be placed independently on a square plane. When perceived as a whole, their visual organiza-

tion depended upon coherence in depth, as well as on the surface. Symmetry was only one of a variety of possibilities for color clusters. While the intention was to satisfy the requirements of a multiplicity of perspectives, the process of problem solving was initiated from a single vantage point. It seemed otherwise impossible to undertake the task. Thus, the Plexiglas framework was studied from its front, where the flat surface of the first plane could be seen, with four behind it in sequence.

8-48

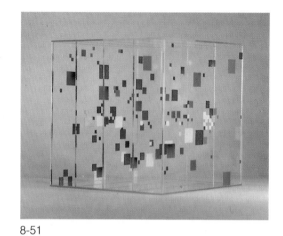

8-51

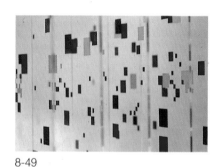

8-49

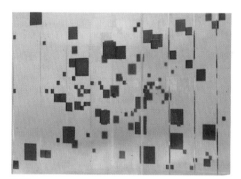

8-50

From this position, all five were experienced spatially. Some students regarded each plane as an entity, making separate two-dimensional patterns against which each, when combined, formed a larger, complex unit.

It was in rare instances that an individual proceeded by placing small chips of colored papers against the planes, intuitively filling the three-dimensional field, as a painter does the canvas. To this person, the challenge was to consider

that each step in the image-building process, would be perceived dynamically, with relation to the next, as one moved around the box. Illustrated is a solution which, as it was photographed from its strongly imaged frontal projection, gradually dispersed as the camera documented its motion around the cube. Features of the constellating effect can be observed, as well as linear sequencing in the grouping of colors through the depth of the field.

8-52

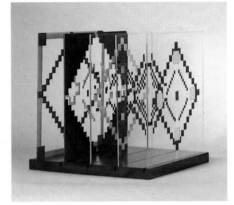

8-53

This sensibility is demonstrated more playfully, in another Plexiglas version of the constellation problem. I suggested that the colored element be considered as a dot floating in space. By leaving open all issues, organization, orientation, and so on, the only unifying principle was that the colored dots be related to the cubic volume.

The inventive result illustrated is constructed of a single sheet of Plexiglas, heated and bent, to form an open box. Then, strung like an abacus, the units of colors were formed as beads on fish lines distributed through the three-dimensional grid of the volume. Since these beads could move freely along the length of string, a fluctuating pattern of colors was obtained, by lifting and rotating the box, and the fixed sequence of hues changed interval and frequency.

8-54

8-55

8-56

8-57

Color and Symmetry

An understanding of the repetition of hues on a plane, is the conception of A.L. Loeb. In his work on color and symmetry,[60] he defines the limited number of groups that is possible between colors, and distinguishes between them numerically. The lucid and logical definition of finite possibility expands, with the prospective use of the perceptual interaction of color as mixtures. Within the 3-3-3 symmetry system of equilateral triangles, four colors were assigned. These were related in their grouping so that two desaturated colors represented middle mixtures between the first brighter and fourth, darker hues.

By grouping these in a plane one can arrange two clusters of large triangles, repeated diagonally through the field, 8-55 or as a broken cluster of 8-56 "bow ties." 8-57 shows the distribution of identical hues with the appearance of a large triangle as "figure," in a figure/ground relationship, or resolved as a series of small, diagonal triangles, overlapped by a delicately transparent large triangle. The complex perceptual possibilities are determined by precise sequence in color, as in all the cases illustrated, the clusters are logically identical.

8-58

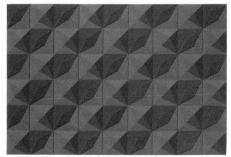

8-59

8-60

It might be noted that the transparency effect in these studies is attained by the use of flat papers, juxtaposed as perceptual intermediaries. When these are aligned in three-dimensional space, as separated planes of color in depth, the flat fields can also appear transparent. In this case the illusion is obtained only when the boundaries of those fields appear to be aligned – that is when the retinally projected image resolves in the illusion of overlapping. While semantically difficult to comprehend, three-dimensional transparencies with flat colors are a visual possibility.

Fourfold Symmetry Adapted

A fourfold rotational pattern was adapted to free arrangements of color. An identical geometric linear pattern underlies a series of paintings, in which flat colors are mixed to appear as transparent areas, by juxtaposition. Complex arrangements and sequences of identical areas can result in very different visual patterns. The fourfold symmetry is concealed, but its rotocenter can be located as the conjunction of four identical isosceles triangles. 8-58, 8-59, 8-60.

8-61

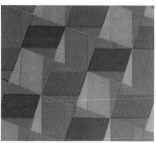

8-62

Three-Dimensional Symmetry

Thinking spatially or three-dimensionally about sequences of colors in dense grids is particularly challenging. Perceptual experiences of this kind require the influence of time and motion. In complex systems color can clarify the position or location of visual elements.

We played with three-dimensional patterns to explore coherent, but complex color orders in space. With students of basic design, I devised a system that combined a three-dimensional grid with tactile strategies of game playing and challenged them to solve the problem. If a cube is used as a field its dimensions can be spatially sectioned by regular intervals, to provide a coherent grid. We used wooden dowels, $^1/_4$ or $^1/_8$ inches in diameter, set into a Masonite peg panel with holes of equivalent size. Each dowel was cut to a 10 inch length, and ten times ten of these formed a linear framework in a 12 inch-square field. The one hundred vertical elements were then painted.

After preliminary exercises with the linear organization of color, four hues were placed as sequences on each dowel, and then on sequences of dowels. Preceeding logically, a group can be colored 1:2:3:4 as a unit, on one stick, to be followed on the next by 2:3:4:1 units, and so on. When such sequences are repeated, adjacent sticks appear to be striped diagonally with clusters of four colors. Each color is applied as a 1-inch interval, or as a proportionally increased size, such as a Fibonacci series. Intervals between colors were marked by white or black to clarify the horizontal grid. Colors were selected as analogous sequences, or discrete, segregating groupings. This choice determined aesthetic or spatial effects. Thus by linear logic, one hundred colored dowels were visually grouped. Since an eight-color sequence was usual at 10-inch length, on each dowel, the colors devised in ascending order for half of its length, might be reversed for the remainder, as a mirror symmetry.

Other conceptual strategies entailed number and system. In one case, a rotational symmetry was devised for a cluster of twenty-five dowels, each identically patterned in color. The rotation of each unit thus presents an alternative face to four sides of the cube in space. Another approach was to conceive of the four faces as equivalent, the frontal, planar conception used in solving the Plexiglas "Mondrian" problem. The possibilities of organizing coherent, solid volumes, were also considered. After initial apprehension about its difficulty, the problem was assiduously resolved by the class.

Since the results can be perceived completely only in time, they are photographed as a series of spatial rotations. To experience them environmentally required that each solution be placed in a neutral and uncluttered visual field. The dowels and their spatial intervals either overlap or align to form transparency effects, relative to their distance from the eye. In a sequence of motion/positions in relation to them, each pattern is visible as a developing and changing order. In an example illustrated, while the lighter blues and grays cluster in perception from one position, they appear to yield to the black and red patterns which begin to emerge from the center of the field, as one moves. A very slight shift in the clockwise rotation will cause these patterns to coalesce, shift, break, and then form newly, elsewhere in the grid. The open feature of the grid provides negative space or interval necessary to the sequencing of the patterns observed; they afford a kind of temporal pacing. The whites or blacks are spatially significant in the horizontal, defining a second grid. Since one becomes rather mesmerized by the color changing the progressive order, these functioned as a reference. Size/distance and other rationalizations of space are subsumed entirely by the organization of color, according to both binocular and monocular observation.

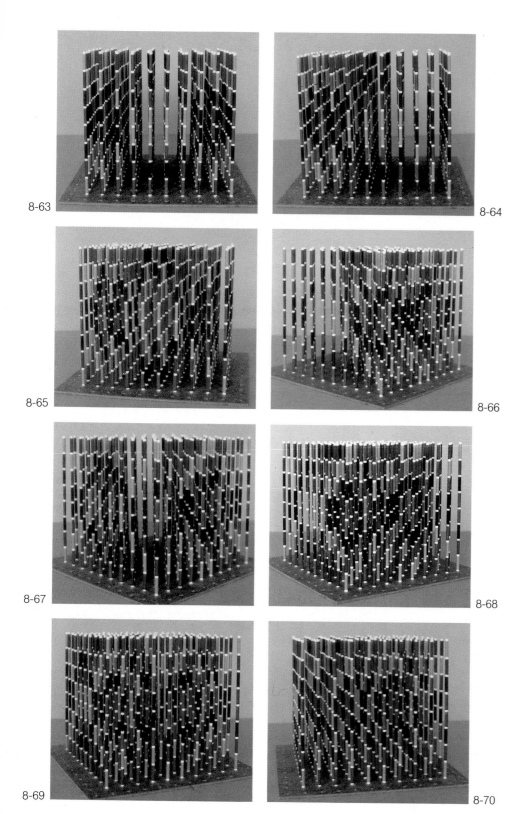

8-63

8-64

8-65

8-66

8-67

8-68

8-69

8-70

This sophisticated version of the color symmetry and constellation effect provides a dynamic, yet coherent strategy for groupings or patterns of color in continuous interaction or rotation. The aesthetic consequences of these orders as open/closed, expanded/contracted, clustered/dispersed, are experienced sequentially in time, and have been resolved uniquely in each case. The linearity, interval, transparency, and constellation of pattern of color here provide the nearest visual equivalence in my experience, to polyphonic music.

8-71

8-72

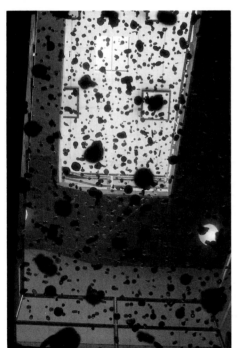

8-73

Pattern and Additive Mixture

The collaborative design and execution of environmental piece was done on the occasion of the completion of a new art facility.[61] A utilitarian stairway, centrally located in the three-store space of the interior at the core of the building, was its major access and circulation. The gray concrete that constituted all walls and surfaces of the stairwell was relieved only by a red tubular steel railing. Since the well was located adjacent to an exterior covered passageway, very little environmental light was available to illuminate this central and significant architectural feature. A clerestory projected past the roof line, bringing natural light from the uppermost level of the vertical shaft of the well. This shifted according to the diurnal path of the sun.

A group of twelve undergraduates undertook to do an analysis of the light, the influence of the existing color and material on the architectural element, and to design and execute a large-scale work.

In the region of upper New York State, sunlight is of short duration, and a gray, overcast condition prevails during the month of January, the project's term. From 7:00 A.M. until 3:30 P.M., illumination rotated in a diagonal path over the four interior walls, diffusely dispersed principally on the third storey, the one most closely adjacent to the source. For the remaining two floors, the heavy closure of gray concrete made the environment and its atmosphere oppressive.

Discussions began with this environmental condition. The group was divided into three groups of individuals, and each team built a small-scale model of the stairwell with a proposed solution, using color. The most obvious one, to paint the walls, was ruled out early, since single, unrelieved surfaces of color would only add to the oppressive sense of containment in the space. Rather, it was felt, something should be done with the negative space itself, the shaft; experienced through all storeys, it offered directional possibilities, and a neutral field for the distribution of color. How to utilize this feature? One group suggested that banners be suspended, or a variety of other smaller colored elements. The problem became how to fix them in space. Other than the railings, there were no three-dimensional elements to which to attach supports.

The geometry of that spatial feature itself suggested the solution. The stair railings provided a consistent and regular structure throughout the entire unit. If the well were to be filled with a linear grid, this could be achieved by attaching lines to and from parallel railings, using them as frames. Since they ran diagonally in the opposing directions of the stairs, oriented parallel to one another around the well, by stringing lines at regular intervals, a surface resulted. This warped surface introduced a curvilinear element into the vertical/horizontal/diagonal architectonic order.

8-74

8-75

Colored elements placed along lengths of string filled the negative interior with color and reflected light, without in any way interfering with the spatial function of the staircase. One of the team's members found an inexpensive source of material, 50-pound bags of small wooden elements, the rejects of toy parts from a nearby factory. Together the group cooperated in spray-painting the ten thousand or so units needed to fill the stairwell.

We accepted the nonuniformity of the elements' shapes as a variant, since they were approximately the same size. This conditioned a decision to introduce a color variant when they were subsequently strung. While each warped plane would be associated with a range of hues, one used yellows and orange-yellows, the next a group of reds and oranges, no single plane would be en-

tirely uniform or repetitious. Thus considered as an entity, a plane represented a color field, but perceived in combinations or overlapped, the individual parts would combine, tending then to be read as an additive mixture. Until the entire assemblage was executed full scale, the visual outcome could not be predicted.

When completed, the interior space of the well seemed filled with an infinity of small colored "dots," and was entirely transformed. Articulated by color, each diagonal plane indicated location on a storey, while from the clerestory, near the roof, looking down, the 120-foot depth of the field seemed indefinite spatially. Colored elements filling its gray air, distributed as a pointillist's dots, had the cumulative effect of dispersing space, and the series of overlapping planes disappeared. From basement level, looking up, the dots lost

their color against the bright field of the clerestory, but appeared as a reversal of the night sky. Here, dark constellations of varying size were dispersed against the brightness. Colored dots marked the difference between lower and upper storey, as well as defining the stairwell space.

Since the resolution of the entire project took one month for analysis, planning, and the rather elaborate execution, refinements were not possible; we realized primary decisions, without modification of ideas in situation. The formal system proved to be fitting and apt, but aspects of scaling in the additive mixture of colors might have been reconsidered. After its brief tenure, absence of the installation was palpably experienced, to the degree that its presence had become integrated as part of the architecture.

9   Color and Light

How do hues behave environmentally, as reflected light? Converted to light, surface colors differ in intensity, by their visual appearance as well as by wavelength. Shifts in the angle of incidence of the sun during the course of the day significantly alter the perceived intensity of individual hues. Highly intense sunlight in Southern California was the environmental condition in which the following experiment with reflected color took place.

9-1

Surface colors appear bright depending on hue and saturation. If they are converted to film, relative intensity changes can be assessed as they are influenced by sunlight. I did a sun study with a group of students, using cylindrical containers of color, grouped in the environment, observed, and visually recorded throughout a diurnal cycle. The sundial, precursor to the clock, served as model and metaphor in the experiment. The heights of the cylinders varied, from 2 to 14 inches, with a circular base 8 inches in diameter. These were placed out-of-doors on a large horizontal plane, where the angle of incidence of the sun played on them in an arc over the course of a day. The interior

of each circular base was painted a different hue, while the curved inner walls of the cylinders were painted white to activate reflectivity from the base to walls. When they were placed in sunlight, with its angle of incidence perpendicular to the base, the reflectivity of all hues was enhanced to maximum. At high noon the empty containers appeared to be brim full of intensely colored light.

The effect of individual colors, enhanced by the scattering of particles of light, differed. We tested ten colors; lemon yellow, cadmium yellow medium, orange, cadmium red medium, alizarin crimson, red-violet, ultramarine blue, blue-green, and cadmium green. Two spectral hues, yellow and red, each were represented in a warm and a cool version, a cool blue and a warm green (containing yellow), their intermediary, blue-green, a single orange and a warm violet were selected, with each color at its maximum saturation.

Cylinder height determined a vertical scale of brightness. With a color applied only on its base, light scattered from the surrounding surfaces was re-

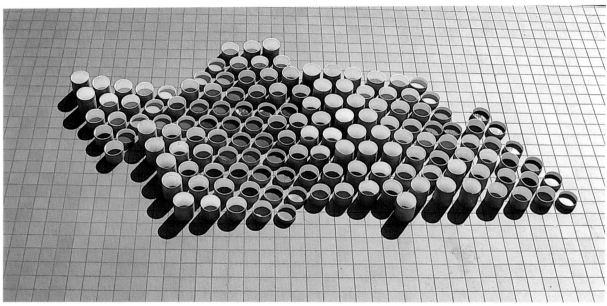

9-2

flected to the degree determined by height, a 4-inch container reflected more saturated hue than one 14 inches high. Color saturation increases with decreased height; the taller the container, the more white of its interior surfaces is added by scattering; in tall containers consequently, colors appeared lighter and less saturated. The experiment was analogous to water glass sound. When tumblers are filled with liquid to different levels, then sounded by tapping their rims, the different volumes will yield changes in pitch. When the amount of liquid relative to the size of the glass is varied, these differences are experienced by the ear as precise tones on a scale.

At the same container height each color, compared at a moment of time, was different in its intensity. At noon, with the sun's angle 90 degrees to the base, all hues were maximally intensified; those relatively brightest were orange-red (cadmium red), orange, and orange-

yellow (cadmium yellow medium), in descending order. The green yellows (lemon yellow) were less intense, followed by green (cadmium green), green-blue, blue and violet, or in a descending order of intensity from long to short wavelength.

Assessed visually, each hue could be reported to reflect its maximum brightness optimally, at a different container height. Once established, the system afforded comparisons. Yellow, for example, attained maximum reflectivity at 4 inches, while warm red was most intense at 8 inches. The 2-inch container was eventually discarded for most colors. It was too short to permit enough scattering, and short enough for the surface color of the base to be visible.

Tested in the studio where they had been painted, all colors had appeared diffuse in the decreased luminosity of the environment, but the cooler green-blues and blues appeared more satu-

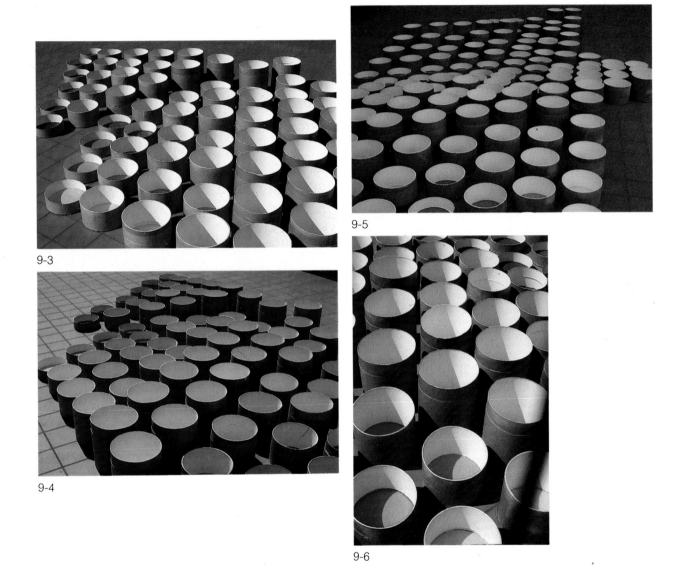

9-3

9-5

9-4

9-6

rated indoors. Outdoors the 150 elements were placed and grouped according to size and color. From 7:00 A.M. until dusk the group was observed and recorded by still photograph and animation camera. Some of the effects are illustrated here. As the sun moved, the moments of rising and descending intensities was recorded as a curvilinear continuum.

9-7

The aesthetic effect of colored light in the containers was magical. At maximum intensities they appeared brimful and palpably colored. As the sun shifted, its path was increasingly defined as a cast shadow against the inner rim of the cylinder. As this widened and deepened, it dispelled the magical illusion by visually relocating the circumference of the container. At the same time, color reflectivity diminished as a function of angular change. When the sun descended at dusk, a rapid transition at this latitude, the entire container remained in shadow, and only residual light remained in its interior to allow barely perceptible color to be distinguished.

The sun study attained more than one objective. We observed the intensity of reflected hues as a function of light, the most direct reading of their physical reality. Intensity was related to wavelength and this was influenced by angular relationship to the source of light.

At 90 degrees to the surfaces, hues were maximally reflected in sunlight, and some appeared most saturated. A vertical scale, the container, added white light to a color, thus increasing lightness in the hue perceived. Spatial expansion and radiation of reflected colors were experienced as a function of wavelength. The longer reds and oranges filled their containers at maximum reflectivity, while greens and blues under the same circumstances floated in the cylinders.

Clustered in groups, colors were actively composed. When juxtaposed they appeared integrated or contrasted, sustained or disrupted as patterns, they vibrated or floated, a constant reference in the environment to the presence of light. Colors directly signified the sun's location by changes in their intensity. In situ the installation recorded angular changes in the rotating path of earth to sun as a diagram of shadows, and a sign of time.

9-8

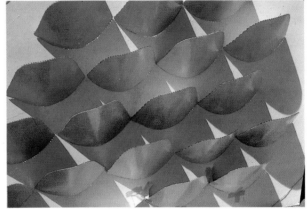

9-9

Modules for Reflected Color

If color is reflected light, then what structure is it that best contains and reflects it? There can be many formal responses to this question. Natural examples abide; feathers, scales, and irridescent cells differentiate wavelengths by unique design strategies, which reflect light as color to the eye.

A class in basic design used paper, glue, and paint to achieve results with simple structural modules. An effective example is illustrated. Heavy, three-ply Strathmore paper was curved, its edges set between two boards and glued. A white, curved plane stood vertically between two horizontal surfaces; in sunlight a chiaro e scuro pattern articulated

the curve and light, reflected from the lower horizontal plane, played over the surface. This effect was enhanced by the addition of color. Brilliant Day-Glo was the student's preference, and she painted the lower horizontal with segments of contrasting hues. When placed in the environment the sun illuminated the horizontal plane, and by reflection cast areas of intense colored light over the curvilinear surface. Angle shifted and the effects changed. Appearing at maximum intensity adjacent to the hori-

zontal source, the color radiated over the curved plane and diminished gradually—an effect analogous to underlighting on the stage.

A "fish scale" was another module devised to work with indirectly reflecting color. Shaped as a gentle curve and painted on its concave inner surface, a number of "scales" were set at an oblique angle to a vertical plane. Progressively overlapping modules reflected the light from their concave surfaces as color cast against the white convex surfaces above. Slanting patterns of bril-

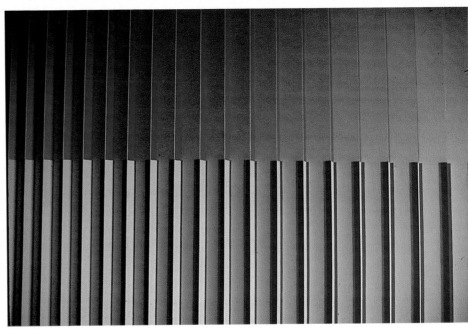

9-11

liantly illuminated color resulted from the proximity of reflecting surface to reflector, a function of sharply oblique angles. The illusion obtained is that of luminous curved surfaces, changing color in sunlight.

Hexagonal units, constructed of paper, folded and glued to one another contiguously, were painted with color on three lower inner surfaces. As the sun's angle shifted, each of the three planes was illuminated at a different time to reflect a color on the remaining three, above.

9-10

A Scale of Shadows

In an exploration of surface in its relationship to light and color, I designed a panel as a geometry which progressively changes size. Constructed of masonite, as a series of triangular sections, "Scale of Shadows" is a relief surface of 90 degree angles. In this piece the 90 degree module was varied by an increase of $^1/_{16}$ inch on one side, while remaining constant on the other. Reading the modules from left to right, the sculptural effect is one of closed to open form. Illuminated from the right, light playing against the sequence of angles cast spreading shadows on the opposite faces. For each section a color between red and blue-violet was mixed at equal intervals, and painted against the angle. I divided the panel in half horizontally, and painted a hue to reflect the light against the right face of

an angle, and placed the same hue against its left side, in the lower half of the field. The visual result was a spreading radiant field of violet sharply divided from a sequence of dark shadows. The penumbra of the shadow was complex in appearance. At the juncture of the angle reflecting color, its complement appeared. While the painted scale ranged from red to blue-violet, the shadow scale read as desaturated yellow to green-gray. Placing the work relative to the studio window, the only light source, I observed changes in its appearance as the light changed. Reflected patterns of colored light radiated, contracted, intensified, diminished; the structure responded to its environment almost as an organism, its geometry made sensuous.

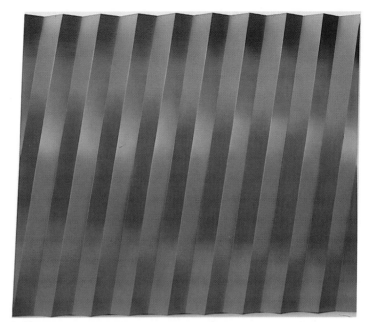

9-12

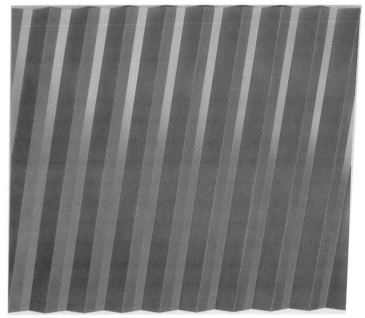

9-13

9-14

10 Visual Education

Visual sensitivity, an aspect of intelligence, is neither a common capacity nor generally the aim of education. Yet over 90 percent of information is obtained visually.

Presumably, the task of basic design is to sensitize vision, but much of its teaching has become academic reiteration of learned formulas. True visual education originates in experience and is rigorous, making demands upon thought processes as well as incurring open reactions.

Children think sensorially, because their visual awareness is keen and they see freshly; their primary mode of experience has not yet been overlaid with rationalizations. Preschool children play and learn directly through their senses. After designing the lucite multiple, I had it available for the use of visitors. At all ages, children were fascinated by it, appreciating its precision and the beauty of its material. Children of four or five explored tactilely as well as visually, freely experimenting with the colored elements. Since the game aspect is creative, play for its own sake, with-

out the issue of "winning" or "losing," the child is challenged to invent strategies, or to discover orders on his own. Preschool children can be particularly logical, and seem to enjoy grading colors according to their sequences, or aligning the blocks' profiles according to their angular relationships. Others respond directly to the aesthetic aspect, the magical appearance of colored light in the clear material. Absorbed for long periods of time, a child will find many solutions to the game of filling a field.

I think that sensory games should be devised and used in early education. Possibly because it is basic, sensory thinking can be accomplished by any child. Organizing patterns and thinking with numbers are closely related activities. For the preschooler, the sensory aspect can be more appealing than counting per se, and represents similar operations, namely, logic and relationship. Early mathematics could remain a concrete issue at the level of game playing for some children who have difficulty with abstraction, until they are ready to make the conceptual leap to number.

In some schools, the use of sensory devices as "visual teaching" aids is merely a rote method. The child who learns by the primary mode, the potential artist, for example, may at this point be alienated from mathematics altogether, when he sees that his conceptual skills are not particularly valued. For it is thinking, and not only the ability to make pictures, which defines this particular form of intelligence. Later in life, the ability to make distinctions between differences in pattern may be important to the biologist, to define the characteristics of a specific virus, to a chemist, in visualizing the complex relationships between elementary components, or to a good mechanic, who can tell, just by looking at it, why an engine will not work. The importance of these perceptual abilities is underestimated and misunderstood at school.

An example of the proficiency with abstraction as it is related to the capacity to distinguish patterns, is recounted by James Watson and Francis Crick, in *The Double Helix*.[62] In constructing the three-dimensional model of the DNA molecule, they were able to clarify the complex relationships between the chemical bases, while visualizing the helical structure. It fit as a puzzle at that stage, confirming the abstract hypothesis.

Visually gifted individuals are often intimidated by mathematics, yet are capable of complex visual operations analogous to those of number. Seeing relationships in patterns, having a sense of how to distribute them, is an attribute of the concrete functioning of color distribution and quantification. In the painter this ability is shared by the abstract artist as well as the pictorial realist. In each case distinctive sensory components are manipulated to make coherent and communicable visual structures.

If at an early age we did not separate the operations of sense so entirely from those of reason, there might be clearer visual thinking, on the one hand, and a better understanding of the strategy of numbers, on the other. These may be similar logical processes, with different terms of expression.

While the environment is three-dimensional and experienced in time, the education of the architect and designer persists as a partite representation of space, in plan, section, elevation. First year architecture students in my class in basic design made a multiple and returned to a thinking level that was directly sensorial, with the result that visual thinking occurred by the operation of concrete components, color, and form.

There are individuals gifted with the capacity to visualize urban patterns. Intuitively understanding the shape of a city, they can find their way in an unfamiliar one almost immediately, by relying upon a few perceptual clues. Others, even with the aid of a map, cannot relate to or orient themselves in unfamiliar surroundings. It may be important to devise visual strategies to inform and generally educate individuals, but it is absolutely essential to the education of the architect and three-dimensional designer.

The model described as a "three-dimensional Mondrian," was a game requiring the invention of pattern by a strategy of placement of colored elements in spaces. In this case, I noted that the capacity to think visually in three dimensions was limited. Some students proceeded linearly, placing each hue independently as a step in a progression. Color clusters or constellations were more often developed as projections or superimpositions of separate two-dimensional patterns in space; very few were coherent, three-dimensional entities.

When the city or urban environment is incoherent, it may be due to human indifference or unplanned obsolescence. More frequently, it occurs as a result of planning by individuals who do not value the capacity to visualize, but think instead statistically and numerically. The sensory link to the environment is lost consequently, and with it human interrelationship. Planned or new cities are sterile, because they are rationalized as functions of economy and utility, and not for their complex human psychological function. In the context of economics, cities are increasingly ad hoc organizations, and chaos replaces in most, what was original order. While the city is the repository of collective memory and human experience, perhaps the most comprehensive representation of culture, and the most complex human artifact; it is increasingly the most problematic to design.

In part this is due to the insignificance of visual education in the culture, accompanied by the trivialization of aesthetics. Increasingly, the concept of the educated individual is one who easily manipulates abstract symbols, instead of one who thinks visually, or behaves humanely. Preschool children are presently being "trained" to acquire reading and numerical skills, prematurely to their ability to utilize them. A confused society underestimates the role of play, imagination, and the exercise of concrete visual sense operations at this level.

I am increasingly aware in teaching that thought processes can proceed as a consequence of alternative strategies, as readily from the concrete to the abstract, as the other way around. It is a matter of context. As I observe the diversity arising as individual responses to the same question, I am convinced that creative thinking has a logic and evolution of its own. Once a problem has been defined, if the path is left open, then independent, progressive conceptual ordering takes place; it may be called invention. We can expect the most originality then, by asking the right questions, and leaving open the issue "how." Originality is not novelty, but a rigorous path followed as a thinking pattern, in response to a sufficiently engaging problem. This is a basic human capacity, whether we choose to call it art, or science, or simply problem solving. It should be the fundamental aim of education, from basis to pinnacle.

Conclusion

While the senses structure experience, the reverse also pertains. Sensation is the result of structure.

Sources of color are structures; the chemical structure of pigments, the crystalline structures that transform light to color, the molecular structure of fluorescence. The structure of the eye itself, reactive to and interactive with the environment, accounts for the experience of reflected light as color.

On another scale, the structural color of natural forms is a matter of architecture. The range and diversity of natural strategies to achieve the effects of color are the result of unique developments. The scales of fish, the feathers of birds, are devices in which relatively few pigments are the basis for the absorption and transmission of light, perceived as the great range, variety and subtlety of color.

Iridescence, that most manifestly beautiful of natural effects, results from a complex and elaborate microstructure which accounts for the magical flash from brilliance to darkness, when a hummingbird shifts and darts. For this effect, each species of iridescent bird manifests a unique microarchitecture in its feathers, which functions essentially to absorb or reflect wavelengths selectively, by reflecting or refracting hues within the complex cell.

As we better understand what surface color is in nature, perhaps a new appreciation will be gained for the inventiveness lavished by evolution on aspects of form that have been regarded as superficial. To what purpose is all this display?

For the designer it provides a new paradigm. Color *is* architecture in nature.

As an aspect of form, surface is as significant, and as fundamentally programmed, as is any other aspect of natural organization.

If design is regarded as the nexus in the relationship between art and science, both attitudes and resources can be influenced by these analogies. The intention need not be to rival natural microengineering for the purpose of display. But seeing an intrinsic relationship pertaining in natural coloration, pattern, and structure, might stimulate their integration by design.

While the language of modernism was influenced by reductive logic, the insights provided by contemporary discoveries of the universe of form might provide a new basis or source for the visual imagination.

At the same time, if this study has challenged reductionism, it reassesses formalism. It suggests that playing with color is a game of invention *and* strategy.

The environment could indeed become metaphoric and meaningful when the designer comes to perceive relationships less limitedly, and is richer and more inclusive in intention.

Notes

[1] Goethe, *Theory of Colors* (London: J. Murray, 1840), Introduction.

[2] S. Edgerton, *The Renaissance Rediscovery of Linear Perspective* (Boston: Basic Books, 1975). The significance of perspective here lies in the rediscovery of classical optics. The function of the eye, by its technological extension, leads to the development of tools of great accuracy. Consequently, navigational implements are developed, which make possible the discovery of the new world, followed by the telescope, etc., and the exponential expansion of human spatial horizons.

[3] F. Allport, *Theories of Perception and Concept of Structure* (New York: Wiley, 1955).

[4] Ibid., p. 120, quotes Koffka.

[5] Ibid., pp. 598–605.

[6] J.J. Gibson, *Perception of the Visual World* (New York): Houghton-Mifflin, 1950). Making the distinction between the visual world, which is external to the human observer, and the visual field, within which these clues interrelate, Gibson redefines the abstract conception of the picture plane of perspectival space.

[7] J.J. Gibson, *The Senses Considered as Perceptual Systems* (Boston: Houghton-Mifflin, 1966). Visual relationship, as expressed by gradients (consistent incremental changes such as size, texture, density) give rise to parallel processing in the cortex.

[8] D. Katz, *The World of Colour* (London: Kegan Paul, 1935). Katz links the loss of surface with the imperceptibility of microstructure or detail. The conversion of a surface to film color results in this condition.

[9] The Liebmann effect is illustrated in a paper study by Josef Albers, using two hues of equal light intensity. When juxtaposed on a plane, the boundary between the two colors is imperceptible.

[10] H. Helson and E. Fehrer, "The Role of Form in Perception," *American Journal of Psychology 44* (1932):79–102.

[11] Defined by Edgar Rubin. The first systematic study of figure/ground—one of his rules says that the enclosed surface tends to be perceived as figure, while the enclosing one will be ground.

[12] J.J. Gibson, *Perception of the Visual World* p. 11.

[13] R.W. Burnham, "The Dependence of Color upon Area," *The American Psychologist 4* (1949):230–231.

[14] F. Birren, *Color Psychology and Color Therapy* (New York: McGraw-Hill, 1950), p. 146.

[15] D. Katz, *The World of Colour* pp. 7–28.

[16] M. Martin, "Film, Surface and Bulky Colors and Their Intermediaries," *American Journal of Psychology 33* (1922):451–480.

[17] D. Katz, *The World of Colour.*

[18] Goethe, *Theory of Colors*, p. 60.

[19] Notes taken by the author in Albers' color course at Yale University in 1952.

[20] J. Albers, *The Interaction of Color* (New Haven, Conn: Yale University, 1963).

[21] Chevreul, *The Principles of Harmony and Contrast of Colors.* (New York: Van Nostrand Reinhold, 1967).

[22] C. Ladd-Franklin, *Color and Color Theories* (New York: Harcourt-Brace, 1929), p. 49: "Hering made it clear that simultaneously induced color was a retinal phenomenon. His theory posits the retinal surface as containing receptor cells sensitive to complementary pairs of colors. The influence of a color on the receptor activates immediately the opposite process. His studies of the after-image and simultaneous light and color induction, place the contrast phenomenon as physiological in character."

[23] Ewald Hering, *Outline of a Theory of the Light Sense* translated by Leo Hurvich and Dorothea Jameson (Cambridge, Mass.: Harvard University, 1964).

24 R. Henneman, "A Photometric Study of the Perception of Object Color," *Archives of Psychology*, Ph.D. Columbia University, New York, 1935.

25 "Sunday Afternoon on the Island of la Grande Jatte." Collection, The Art Institute of Chicago.

26 Leo Hurvich and Dorothea Jameson, "From Contrast to Assimilation; in Art and in the Eye," Pergamon, *Leonardo 8* (1975):125–131. The effect of additive mixtures by their assimilation in the eye is described with respect to its occurrence in the work of the pointillist painters, and scientifically, in the optical response.

27 E. Hering, *Theory of the Light Sense.*

28 Hermann von Helmholtz, *Treatise on Psychological Optics* (New York: Dover, 1962.

29 Edwin Land, "The Retinex," *The American Scientist 52*, 10.2 (1964).

30 C. Ladd-Franklin, *Color and Color Theories.*

31 E.T. Hall, *The Hidden Dimension* (New York: Doubleday, 1966).

32 A.H. Munsell, *A Color Notation*, 12th ed. (Baltimore: Munsell Color Company, 1975).

33 W. Ostwald, *The Color Primer* (New York: Van Nostrand Reinhold, 1969).

34 P. Bouma, *The Physical Aspects of Color* (The Netherlands: Eindhoven Phillips Gloeilampenfabrieken, 1947).

35 The Weber-Fechner law—The arithmetic appearance of light gradation is due to a geometric progression of the stimulus.

36 Albers, *Interaction of Color.*

37 The term *aerial perspective* denotes the optical change in surfaces seen at great distances. Their distinctiveness as structures diminishes and at the same time all colors are influenced by the appearance of blue. The term is found in Leonardo's notebooks, and in Alberti's *Treatise on Painting.*

38 Koschmieder defined the theory of visual range in 1924, from R.A.R. Tricker, *Introduction to Meteorological Optics* (New York: American Elsevier, 1970).

39 Fechner's psychophysical law. As the distance from the stimulus increases relative to the eye, the contrast between adjacent areas decreases in intensity.

40 Albers, *Interaction of Color.*

41 Except where otherwise noted, Color-aid paper is used for the experiments described in this chapter, and in those on form. The terms "tints" and "shades" are the manufacturer's description for values.

42 D. Katz, *The World of Colour.*

43 Albers, *Interaction of Color.*

44 L. Swirnoff, Spatial Aspects of Color. M.F.A. thesis, Yale University, 1956.

45 D. Katz, *The World of Colour*, p. 224.

46 Gwen Putzig, an undergraduate at Skidmore College.

47 Allport, *Theories of Perception and Concept of Structure.* "Holway and Boring found that when all distance clues are eliminated except visual angle, apparent size varies with distance about as visual angle does; in other words, the size constancy of the perceived object is lost."

48 Leonardo da Vinci, *Notebooks*: "The shape of a body cannot be accurately perceived when it is bounded by a color similar to itself."

49 Ralph Knowles, *Energy and Form: an Ecological Approach to Urban Growth* (Cambridge, Mass. and London: MIT Press, 1974).

50 P. Mondrian, *Plastic Art and Pure Plastic Art.*

51 Hans Jaffé, The Erasmus Lecture at Fogg Museum, Harvard University, 1980.

52 Carlo Pedretti in his Commentary to Jean Paul Richter's *The Literary Works of Leonardo da Vinci*, 1977.

53 Albers made the distinction between what appears to be a local, retinal phenomenon, and an effect which takes place further in the visual system, which he termed, "perceptual." He much preferred this term to "optical," adopted by the proponents of "Op" Art of the sixties, who attribute their work to his influence. "Perceptual" includes the influence of visual experience beyond the visual stimulus.
In this instance, I think that the resolution of the effect lies in the propensity of the eye and brain to make ratios of the stimulus of light and to reduce the variables of hue.

54 R. Arnheim, *The Power of the Center* (Berkeley: University of California Press, 1982). The psychological significance of the center is discussed here, in the composition of paintings and in design.

55 P. Stevens, *Patterns in Nature.* Those morphological features which repeat in nature—meanders, spirals, branching systems, fluid dynamics—are explained as the properties of space-filling. Diverse forms are thus linked by their pattern-making propensites.

56 A. Portmann, *Animal Forms and Patterns* (New York Schocken Books, 1952). The Swiss zoologist maintains that surface patterning is a significant aspect of organic form. He argues that patterns exist, not solely for functional reasons, but as the means by which organisms are differentiated. For example, he states that the interiors of animals are organically similar; it is only on their surfaces that species are identified, and display individual characteristics. He places therefore, an intrinsic value on the surface.

57 Leonardo considered the mastery of chiaro e scuro to be "the source of excellence in painting," and he predicted that "its science would engender great discourse." Pedretti, commentary on Richter, "The Literary Works of Leonardo da Vinci."

58 Allport, *Theories of Perception and Concept of Structure.*

59 Swirnoff, *Spatial Aspects of Color.*

60 A.L. Loeb, *Color and Symmetry* (New York: Wiley, 1971). In collaboration with the author in a section of "Dimensional Color," at Harvard.

61 The Art Building at Skidmore College, Saratoga Springs, N.Y., 1980.

62 James Watson and Francis Crick, The Double Helix (New York: Norton, 1980) In the personal and anecdotal account of the Nobel-winning discovery of the structure of DNA, the ability to play and imagine is demonstrated on a highly sophisticated level.

Bibliography

Albers, Josef. *The Interaction of Color*. New Haven Conn.: Yale University Press, 1963.

Allport, Floyd. *Theories of Perception and Concept of Structure*. New York: Wiley, 1955.

Arnheim, Rudolf. *Art and Visual Perception*. Berkeley: University of California Press, 1954.

– *The Power of the Center*. Berkeley: University of California Press, 1982.

Bezold, W. Von. *The Theory of Color*. Boston: Prang, 1876.

Birren, Faber. *New Horizons in Color*. New York: Reinhold, 1955.

Boring, E.G. *A History of Experimental Psychology*. New York: Appleton-Century-Crofts, 1950

Bouma, Pieter. *The Physical Aspects of Color*. Eindhoven, The Netherlands: Philips Gloeilampenfabrieken, 1947.

Burnham, Robert. "The Dependence of Color upon Area." *American Psychologist* (July 1949): 230–231.

Chevreul. *The Principles of Harmony and Contrast of Colors*. New York: Van Nostrand Reinhold, 1967.

Edgerton, Samuel. *The Renaissance Rediscovery of Linear Perspective*. Boston: Basic Books, 1975.

Evans, Ralph. *An Introduction to Color*. New York: Wiley, 1948.

Fehrer, H. and E. Helsen. "The Role of Form in Perception." *American Journal of Psychology 44* (1932): 79–102.

Friedberg, Mildred. *De Stijl: 1917–31 Visions of Utopia*. Walker Art Center Catalogue. New York: Abbeville Press, 1982.

Fry, G.A. and V.M. Robertson. "Alleged Effects of Figure/Ground upon Hue and Brilliance." *American Journal of Psychology 47* (1935): 424–435

Gibson, J.J. *The Perception of the Visual World*. Boston: Houghton Mifflin, 1950.

– *The Senses Considered as Perceptual Systems*. Boston: Houghton Mifflin, 1966.

Gilchrist, A.L. "The Perception of Surface Blacks and Whites." *Scientific American 240*, no. 3 (March 1970).

Goethe. *Theory of Colors*. London: J. Murray, 1840.

Gombrich, E.H. and R.L. Gregory. *Illusion in Art and Nature*. New York: Scribner, 1973.

Hall, E.T. *The Hidden Dimension*. New York: Doubleday, 1966.

Helmholtz, H. Von. *Treatise on Physiological Optics,* edited by J. Southall. New York: Dover, 1962

Henneman, Richard. "A Photometric Study of the Perception of Object Color." *Archives of Psychology*. Ph.D. dissertation. Columbia University, 1935.

Hering, Ewald. *Spatial Sense and Movements of the Eye*. Baltimore: American Academy of Optometry, 1942.

– *Outline of a Theory of the Light Sense*, translated by Leo Hurvich and Dorothea Jameson. Cambridge, Mass.: Harvard University Press, 1964.

Hurvich, Leo and Dorothea Jameson. "From Contrast to Assimilation; in Art and in the Eye." Pergamon, *Leonardo 8* (1975): 125–131.

Ittleson, W. *The Ames Demonstrations in Perception.* Princeton, N.J.: Princeton University Press, 1952.

Katz, David. *The World of Colour.* London: Kegan Paul, 1935. New York: Johnson Reprint Co., 1970.

Koffka, Kurt. *The Principles of Gestalt Psychology.* New York: Harbinger Books, Harcourt, Brace and World, 1935.

Knowles, Ralph. *Energy and Form: an Ecological Approach to Urban Growth.* Cambridge, Mass.: MIT Press, 1974.

Krakauer, L.J. "Computer Analysis of Visual Properties of Curved Objects." Cambridge, Mass. Project MAC, 1971.

Ladd-Franklin, C. *Color and Color Theories.* New York: Harcourt Brace, 1929.

Land, Edwin. "The Retinex." *The American Scientist 52,* no. 2 (1964).

Leonardo da Vinci. *Selections from the Notebooks.* Oxford: Oxford University Press, 1952.

Loeb, A.L. *Color and Symmetry.* New York: Wiley, 1971; Krieger, Huntington, N.Y., 1978.

Luckiesh, Matthew. *Color and Its Application.* New York: D. Van Nostrand, 1915.

Martin, Mabel. "Film, Surface and Bulky Colors and Their Intermediaries." *American Journal of Psychology 33* (1922): 451–480.

Matthaei, Rupert. *Goethe's Color Theory.* New York: Van Nostrand Reinhold, 1971.

Mondrian, Piet. *Plastic Art and Pure Plastic Art.* New York: Documents of Modern Art, Wittenborn, 1947.

Mumford, Lewis. *The City in History.* New York: Harcourt, 1961.

Munsell, A.H. *A Grammar of Color.* New York: Van Nostrand Reinhold, 1969.

– *A Color Notation*, 12th ed. Baltimore: Munsell Color Co., 1975.

Ostwald, Wilhelm. *The Color Primer.* New York: Van Nostrand Reinhold, 1969.

Pedretti, Carlo. Commentary to Jean Paul Richter's *The Literary Works of Leonardo da Vinci.* 1977.

Portmann, Adolph. *Animal Forms and Patterns.* New York: Schocken, 1952.

Rock, Irvin. *Perception.* New York: Scientific American Library, 1984.

Rubin, E. "Visuell wahrgenommene wirkliche Bewegungen." *Zeitschrift für Psychologie,* 1927, *103,* 389–392

Solon, Leon. "Polychromy." *The Architectural Record.* New York, 1924.

Stevens, Peter. *Patterns in Nature.* Boston: Little Brown, 1974.

Swirnoff, Lois. *Spatial Aspects of Color.* M.F.A. thesis, Yale University, New Haven, Conn. 1956.

– "Experiments on the Interaction of Color and Form." *Leonardo 9* (1974): 191–195.

Thomson, D'Arcy. *On Growth and Form.* London: University Press, 1961.

Tricker, R.A.R. *Introduction to Meteorological Optics.* New York: American Elsevier, 1970.

Vinacke, W. "The Discrimination of Color and Form at Levels of Illumination Below Conscious Awareness." *Archives of Psychology* (1942): 5–53.

Wallach, H. "Brightness Constancy and the Nature of Achromatic Colors." *Journal of Experimental Psychology 38* (1948): 310–324.

Watson, James and Francis Crick. *The Double Helix.* New York: Norton, 1980.

Werner, H. "Studies on Contour: I. Quantitative Analysis." *American Journal of Psychology 47* (1935): 40–64.

List of Illustrations

Chapter	Illustration	Title	Credit
Chapter 1	1-1	Rome, Italy	Lois Swirnoff
	1-2	Old Town, Stockholm	,,
	1-3	Venice, Italy	,,
	1-4	Venice, Italy	,,
	1-5	Venice, Italy	,,
	1-6	Jalapa, Vera Cruz	,,
	1-7	Oaxaca, Mexico	,,
	1-8	Puebla, Mexico	,,
	1-9	Parma, Italy	,,
	1-10	Parma, Italy	,,
	1-11	Venice, Italy	,,
	1-12	Old Town, Stockholm	,,
	1-13	Jalapa, Vera Cruz	,,
	1-14	Grand Canal, Venice	,,
	1-15	Old Town, Stockholm	,,
	1-16	Bethlehem, Israel	,,
	1-17	Jerusalem, Israel	,,
	1-18	Jalapa, Vera Cruz	,,
	1-19	Parma, Italy	,,
	1-20	Venice, Italy	,,
	1-21	Cambridge, England	,,
	1-22	San Miniato, Florence, Italy	courtesy of Sussman/Prejza
	1-23	Orvieto Cathedral, Italy	courtesy of Tina Beebe
	1-24	Orvieto Cathedral, interior	
	1-25	Majid-I-Shah, Isfahan	courtesy of A. Ricks

List of Illustrations (continued)

Chapter	Illustration	Title	Credit
Chapter 1	1-26	Majid-I-Shah, Isfahan	A. Ricks
	1-27	Olympics, 1984, Los Angeles	courtesy of Annette del Zoppo
	1-28	Sussman/Prejza Co. and	,,
	1-29	Jerde Associates	,,
Chapter 2	2-1	Georges Seurat, Sunday Afternoon on the Island of La Grande Jatte, 1884–86, oil on canvas, 207.6 × 308.0 cm	Helen Birch Bartlett Memorial Collection, 1926.224. ©1988 The Art Institute of Chicago. All Rights Reserved.
	2-2	Munsell System	
	2-3	the visible spectrum	
	2-4	Oscar Claude Monet, Haystack at Sunset near Giverny, oil on canvas, 73.3 × 92.6 cm.	Juliana Cheney Edwards Collection, Museum of Fine Arts, Boston
	2-5	Color study in paper	undergraduate in Design, UCLA
	2-6	Mark Rothko, Number 10, 1950, oil on canvas, 7′6 3/8″ × 57 1/8″	Collection, The Museum of Modern Art, New York. Gift of Philip Johnson
	2-7	Fan K'uan, Travelling among Streams and Mountains Sung Dynasty 960–1279	Collection of the National Palace Museum Taipei, Taiwan, Republic of China
	2-8	California Desert	Lois Swirnoff
	2-9	Grand Canyon, Arizona	,,
Chapter 3	3-1	Window Problem: a series of single colors seen through a gray window; orange-/red hue	Allison Kume Beth Beymer undergraduates in Design, UCLA
	3-2	Blue hue	,,
	3-3	Red hue	,,
	3-4	Desaturated light blue	,,
	3-5	Dark blue	,,
	3-6	Yellow/white window	,,
	3-7	Yellow/gray window	,,

List of Illustrations (continued)

Chapter	Illustration	Title	Credit
Chapter 3	3-8	Light violet, gray window Series of two colors through a gray window	undergraduates in Design, UCLA
	3-9	Bright adjacent to dark color	,,
	3-10	Blue and green appear to share the same spatial plane (Liebmann effect)	,,
	3-11	Vibration/red and green	,,
	3-12	Sequence of 4 flat colors appears as 4 overlapping transparent squares	,,
	3-13	Sequence of 4 colors appears in reverse order	,,
	3-14	Sequence of 4 colors appears as a longitudinal space	,,
	3-15	Sequence of intermediaries between orange and violet	Thomas M. Carr undergraduate in Design, UCLA
	3-16	5 flat grays appear volumetric	
	3-17	the grays in reverse order	
	3-18	3 spatially separated flat colors appear to intersect	Lois Swirnoff, *Spatial Aspects of Color*, M.F.A. thesis, Yale University, 1956
	3-19	3 spatially separated flat colors appear as 2 overlapping planes	,,
	3-20	Flat circles appear volumetric	undergraduate in Design, UCLA
	3-21	2-dimensional study of 3 flat colors which appear as 2 intersecting	,,
	3-22	3-dimensional study; 3 flat colors in space appear as a color stereopsis	,,
	3-23	3 flat colors in space appear as a zig-zag	,,

List of Illustrations (continued)

List of Illustrations (continued)

Chapter	Illustration	Title	Credit
Chapter 6	6-1	Mexican parrot	Lois Swirnoff
	6-2	Moth on pine bark	courtesy of Alexander B. Klots
	6-3	2 frogs	
	6-4	Oriental bird	Lois Swirnoff
	6-5	Surgeonfish	courtesy of H. Kacher
	6-6	Young Imperial Angelfish	,,
	6-7	Red Sea	Lois Swirnoff
	6-8	Angelfish	courtesy of Carl Roessler
	6-9	Triggerfish	courtesy of Gene Wolfsheimer
	6-10	Autumn at the Muddy River Park, Brookline, Massachusetts	Lois Swirnoff
	6-11	,,	,,
	6-12	Dying oak leaf	,,
	6-13	Anza Borrego, California	,,
	6-14	Desert blossom, Anza Borrego	,,
	6-15	Judean Desert, Israel	,,
	6-16	Tendril	Lois Swirnoff
	6-17	Shell	courtesy of Robert Oberhand
	6-18	Zebras	courtesy of Martin Johnson The American Museum of Natural History, New York
	6-19	Tidal Pool, Martha's Vineyard, Massachusetts	Lois Swirnoff
	6-20		
	6-21		
	6-22		
Chapter 7	7-1	Leonardo da Vinci, The Madonna, Infant and St. Anne; chalk on paper	collection of The Royal Academy of Art, London
	7-2	Jean Auguste Ingres, Study for "Raphael and the Fornarina"	The Metropolitan Museum of Art, Robert Lehman Collection, 1975 (1975.1.646)

List of Illustrations (continued)

Chapter	Illustration	Title	Credit
Chapter 7	7-21	Free study	undergraduate in Design, UCLA
	7-22	Free study	Leah Lee
	7-23	Free study, pyramid, lateral view	undergraduate in Design, UCLA
	7-24	Free study	,,
	7-25	8 colors appear as 4 areas, flattening the pyramid	Kimberly Allen undergraduate in Design, UCLA
	7-26	16 colors appear to define 4 pyramids	Mark Van Slyke undergraduate in Design, UCLA
	7-27	Free study	,,
	7-28	Free study, pyramid/lateral view	Lauren Fischer undergraduate in Design, UCLA
	7-29	Free study	,,
	7-30	Free study	Iris Wilfong undergraduate in Design, UCLA
	7-31	Free study	undergraduate in Design, UCLA
	7-32	Free study	Sam Kim undergraduate in Design, UCLA
	7-33	Free study, pyramid/lateral view	undergraduate in Design, UCLA
	7-34	Free study	,,
	7-35	Free study	Christopher Roberts undergraduate in Design, UCLA
	7-36	Free study	undergraduates in Design, UCLA
	7-37	Free study	,,
	7-38	Free study, pyramid/lateral view	undergraduate in Art, Skidmore College
	7-39	Free study	,,
	7-40	Free study	,,
	7-41	Free study	Joanne Nicholson undergraduate in Design, UCLA
	7-42	Free study	Brenda Lawler undergraduate in Art, Skidmore College

List of Illustrations (continued)

Chapter	Illustration	Title	Credit
Chapter 8	8-18	Multiple: Mailing tubes 144 units	Walter Bender undergraduate in VES, Harvard University
	8-19	Unit as module	
	8-20	25 pyramids comprise a Wall 12 × 12 ft.	Donald Koblintz, installation at Carpenter Center, Harvard University
	8-21	Detail of wall illuminated by incandescent light	
	8-22	A single pyramid photographed in rotation comprises a Wall 6 × 6 ft.	Thomas M. Carr undergraduate in Design, UCLA
	8-23	Modular fan-shaped unit as parallelogram	Peter Chandler undergraduate in Art, Skidmore College
	8-24	Module in hexagonal formation	
	8-25	Hexagon, rotated in light	,,
	8-26	,,	,,
	8-27	Module in close-packed configurations	,,
	8-28	,,	,,
	8-29	,,	,,
	8-30	Cube module	William Nesto, Douglas Payne, Andrea Kirsch, undergraduate students in VES, Harvard University
	8-31	Module in formation	
	8-32	,,	
	8-33	Cube module, stacked	
	8-34	Bead multiple	undergraduates in Design, UCLA
	8-35	Bead multiple	,,
	8-36	Bead multiple	,,
	8-37	Bead multiple	,,
	8-38	Sticker multiple	Eun-Hee Kim, UCLA
	8-39	Sticker multiple	Marco Perella, UCLA
	8-40	Computer generated pattern	Jim Squires graduate student in Design, UCLA
	8-41	Color change	

List of Illustrations (continued)

Chapter	Illustration	Title	Credit
Chapter 8	8-65	Model in rotation	Heidi Freeman
	8-66	,,	,,
	8-67	,,	,,
	8-68	,,	,,
	8-69	,,	,,
	8-70	,,	,,
	8-71	Environmental installation temporary: 4 storey stairwell, Art building, Skidmore College: acrylic on wood, strung on fish line; 10,000 elements	Skidmore undergraduate students in Art: Amy Williams Nancy Betke Jennifer Charbonnier Nancy Crane Robin Hunt Elizabeth Klein Betsy Linzer Susan Regis Donna Sangenito Linda Weise
	8-72	,,	
	8-73	,,	
	8-74	,,	
	8-75	,,	
Chapter 9	9-1	Color as reflected light: acrylic on cylindrical containers	Undergraduates in Design, UCLA
	9-2	Assembled group at high noon	
	9-3	Shadow pattern with 45 degree angle	
	9-4	Reflectivity of hues at different cylinder heights noon	
	9-5	Dusk	
	9-6	Film color effect dispelled by cast shadow	
	9-7	Maximum reflectivity of hues with angle of incidence of sun 90 degrees to base	

List of Illustrations (continued)

Chapter	Illustration	Title	Credit
Chapter 9	9-8	Reflected color on curved surface, Day Glo on Strathmore paper	undergraduates in Design, UCLA
	9-9	Reflected color, "fish scale", acrylic on paper	,,
	9-10	Reflected color, hexagonal module, acrylic on paper	,,
	9-11	"Scale of Shadows" 48″ × 72″ × 1 1/2″ acrylic on masonite	Lois Swirnoff collection of the artist
	9-12	"Transformation" 22″ × 26″ × 3/4″ acrylic on aluminum	Lois Swirnoff collection of the artist
	9-13	"Desert Light" 22″ × 26″ × 3/4″	Lois Swirnoff collection of Alice March
	9-14	"Soft Symmetry" 48″ × 54″ × 1 1/2″, acrylic on aluminum	Lois Swirnoff, collection of the artist

Index

Numbers in parentheses refer to the number of the Notes (pp. 139–140).
Numbers set in *italics* indicate the page on which the complete literature citation is given.

Dimensional Color

by Lois Swirnoff

A *Pro Scientia Viva* Title

A volume in the *Design Science Collection*
Arthur L. Loeb, Series Editor

The perception of forms is affected by their context. When the dimension of color is added, these effects can be complex and paradoxical.

Josef Albers' studies in two dimensions explored the interaction of color in area and field, greatly enhancing the painter's understanding of color formalism.

Lois Swirnoff's experiments in *Dimensional Color* begin with the premise that a color changes when it is influenced by its background, but she extends the frame of reference to the environment itself. Color and surface, color as space, color as a primary aspect of visual organization, are visual issues analyzed in the context of three dimensions.

Swirnoff has developed a grammar of relationships between color and form. She challenges the designer to regard color as intrinsic to the design process. Complex form will result, offering a logical alternative to the prevailing eclecticism and revivalism of post-modernism.

"This book offers an imaginative and original exploration of the effect of color and brightness on three-dimensional space and volume. As such it is of immediate interest to designers, architects, and to some extent, sculptors. It offers a beautiful opportunity for the training of sensitivity and visual inventiveness in the studio applied to materials that go far beyond those introduced by Albers."

— RUDOLF ARNHEIM
University of Michigan

"*Dimensional Color* will significantly add to the syntax and grammar of color in the designed world. As a painter very much involved with acute color perception and as one who over these many years both trained and collaborated with architects, I know just how valuable these published experiments will be both in the school studios and in professional offices."

— ROBERT SLUTZKY
The Cooper Union Art School

"Very stimulating and memorable … Ms. Swirnoff has clearly given a rational base, a predicament for the manipulation of colors to produce three-dimensional illusions … I can certainly recommend it as a textbook for courses that deal with color and its application in architecture and design."

— Friedrich St. Florian
Rhode Island School of Design

"Lois Swirnoff works from the basic premise that form, color, and texture are insepa- rable. In her book she demonstrates experimentally how the observed dimensions of physical forms will vary with color-context as well as the angle between incident light, the surface, and the observer. Her work is of fundamental interest to anyone who wishes to design spaces within given physical constraints, whether these spaces are urban places, interiors, or theatrical settings."

— Arthur L. Loeb
Series Editor

About the Author

Painter, photographer, and visual logician, Lois Swirnoff's career spans over two decades as well as the east and west coasts of the United States.

Currently on the Faculty of Design at University of California at Los Angeles, she has taught at Harvard University, at Skidmore and Wellesley Colleges, and at Universi- ty of Southern California.

A graduate of The Cooper Union Art School in New York, Swirnoff earned her B.F.A. and M.F.A. at Yale University where she studied with Josef Albers. She was a Fulbright Scholar in Painting to Italy, as honorary representative of New York State, a Fellow in Painting of the Mary I. Bunting Institute of Radclife College, and a Fellow at Yaddo.

Swirnoff's work is included in public and private collections, and in addition to exhibitions, her articles on color and the environment can be found in *Leonardo* and *The Environmentalist*.